REVISE EDEXCEL GCSE
Science

REVISION GUIDE
Higher

Series Consultant: Harry Smith
Series Editor: Penny Johnson

Authors: Penny Johnson, Sue Kearsey,
Damian Riddle

THE REVISE EDEXCEL SERIES
Available in print or online

Online editions for all titles in the Revise Edexcel series are available Autumn 2012.

Presented on our ActiveLearn platform, you can view the full book and customise it by adding notes, comments and weblinks.

Print editions

Science Revision Guide Higher	9781446902615
Science Revision Workbook Higher	9781446902622

Online editions

Science Revision Guide Higher	9781446904893
Science Revision Workbook Higher	9781446904916

Print and online editions are also available for Science (Foundation), Additional Science (Higher and Foundation) and Extension Units.

To find out more visit:
www.pearsonschools.co.uk/edexcelgcsesciencerevision

ALWAYS LEARNING

PEARSON

Contents

A small bit of small print

Target grade ranges are quoted in this book for some of the questions. Students targeting this grade range should be aiming to get most of the marks available. Students targeting a higher grade should be aiming to get all of the marks available.

Edexcel publishes Sample Assessment Material and the Specification on its website. This is the official content and this book should be used in conjunction with it. The questions in *Now try this* have been written to help you practise every topic in the book. Remember: the real exam questions may not look like this.

Classification

Classification groups

Classification means grouping things by their features or characteristics. The more characteristics that are used to group similar organisms together, the more reliable is the classification.

Small groups that are similar can be grouped into larger groups. Kingdoms are the largest groups in classifying organisms, and species are the smallest.

A mnemonic like this can help you remember the groups in the right order: Keep Pond Clean Or Frogs Get Sick.

Kingdom	Phylum	Class	Order	Family	Genus	Species
Group of similar phyla e.g. Animalia	Group of similar classes e.g. Chordata	Group of similar orders e.g. Mammalia	Group of similar families e.g. Carnivora	Group of similar genera e.g. Canidae (dog family)	Group of similar species e.g. Canis	Organisms that have most characteristics in common e.g. Canis lupus (domesticated dog)

The five kingdoms of organisms

1 Animalia (animals)
- multicellular (body made of many cells)
- have no cell walls
- no chlorophyll in cells
- feed heterotrophically (eat other organisms)

3 Fungi
- multicellular
- have cell walls
- no chlorophyll
- feed saprophytically (digest food outside the body)

2 Plantae (plants)
- multicellular
- have cell walls
- have chlorophyll
- feed autotrophically (make own food)

4 Protoctista
- mostly unicellular (body is a single cell)
- nucleus in cell

5 Prokaryotae (mostly bacteria)
- unicellular
- no nucleus in cell

Worked example

Viruses are not classified in a kingdom. Explain why.

They are not classified because most scientists think that viruses are not living organisms. This is because:
- They have to use other cells to reproduce.
- They show no other life processes (for example, growth).

Now try this

1. *Paramecium* is a microscopic pond organism. Identify which kingdom it belongs in and explain your answer. **(2 marks)**

target **D-C**

nucleus

2. The common frog and viper (snake) are grouped in the phylum Chordata. The viper and chameleon are grouped in the class Reptilia. Identify whether the frog or the chameleon shares more characteristics with the viper. Explain your answer. **(2 marks)**

target **D-B**

Vertebrates and invertebrates

The phylum Chordata contains animals that have a supporting rod running the length of their body. Many animals in the Chordata are vertebrates (animals with backbones).

Vertebrate groups

Scientists sort vertebrates into five main groups.

oxygen absorbtion
inside blue area: use gills
inside green area: use lungs
(note: young amphibians use gills, but adult amphibians use lungs)

reproduction
inside pink area: young born live (viviparous)
outside pink area: lay eggs (oviparous)

fertilisation
inside yellow area:
inside female's body (internal)
outside yellow area:
outside female's body (external)

thermoregulation
inside purple area: control own body temperature (homeotherm)
outside purple area: body temperature varies with environment temperature (poikilotherm)

Worked example

Bats and birds are vertebrates with wings, but they are classified in different vertebrate groups. Use these examples to help explain why it is sometimes difficult to classify vertebrates.

Bats and birds both have wings and can fly, so if you only looked at these characteristics you would put them in the same vertebrate group. Both bats and birds absorb oxygen through their lungs and they are homeothermic. But other characteristics show that they are very different and should be classified in different groups. Bats are mammals because they give birth to live young. Birds have feathers and lay hard-shelled eggs (this is called oviparous).

Organisms that live in similar ways often have similar characteristics as a result of adaptation to their environment. It can be easy to confuse these characteristics with those shared as a result of evolution.

Now try this

target **D-C**

1. Birds and reptiles belong to two different classes of vertebrates.
   (a) Give two key characteristics that the birds and reptiles share.   **(2 marks)**
   (b) Give one characteristic that is used to separate the reptile and bird classes.   **(1 mark)**

target **C-A**

2. Sharks and dolphins are marine predators that look similar. Sharks are classified as fish and dolphins as mammals. Explain why they are not classified in the same class even though they look similar.   **(3 marks)**

dolphin   shark

Species

Defining 'species'

A species is a group of organisms that can breed with each other and produce fertile offspring.

A few species can interbreed (breed with other species) to produce hybrid offspring, but often these hybrids are sterile (cannot produce offspring).

Problem 1: Some species reproduce asexually from parts of their body or by dividing in two. If we do not see two organisms breeding together we cannot be sure if they are the same species.

Problem 2: Some organisms from closely related species can interbreed and produce hybrid offspring that *are* fertile.

Constructing and using keys

A key based on obvious characteristics can help identify different species. This is a key for some big cats.

1	Has no spots?	go to 2
	Has a spotted coat?	go to 4
2	Has stripes?	tiger
	Has no stripes?	go to 3
3	Has ring of dark fur round nose and mouth?	puma
	Has no ring of dark fur?	lion
4	Has simple spots of same size all over?	cheetah
	Has rosettes (patterns of spots)?	go to 5
5	Rosettes have small dots inside?	jaguar
	Rosettes have no dots inside?	leopard

To construct a key, start by separating the organisms into groups using obvious **characteristics**. Continue separating each group into smaller groups by their characteristics until you have only one organism in each group.

Worked example

Explain how scientists share and validate new discoveries.

They write a paper and send it to a scientific journal. The paper is checked first by other scientists to see that it is good enough to publish. This is called a peer review. They could also present their discovery to other scientists at a conference. When other scientists hear about the discovery, they will check it by repeating the experiments. If they get the same results, then the discovery is validated.

Now try this

target D-B

1. Explain the meaning of these words:
   (a) fertile
   (b) hybrid
   (c) species. **(3 marks)**

2. Construct a key to identify these four animals. **(4 marks)**

Binomial classification

Every species has a unique binomial name, for example the African lion is *Panthera leo*.

Panthera leo

genus name: shared with very similar species

species name: unique to African lion

The leopard (*Panthera pardus*) and the tiger (*Panthera tigris*) are closely related to the lion so have the same genus name. The cheetah (*Acinonyx jubatus*) and snow leopard (*Uncia uncia*) are NOT closely related to the lion, so are placed in different genera.

Binomial names are useful because

- other people know exactly which species you mean
- you can see from the genus which species are very closely related
- it helps us identify which environments contain the fewest species (low biodiversity) that are most at risk of extinction and need the most protection and conservation.

Classification complications

Classifying an organism within a particular species may not be easy.

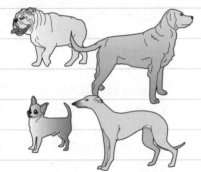

Dogs show a wide range of variation but all belong to one species.

characteristics of pochard

characteristics of ferruginous duck

hybrid duck
head shape and white tail flash from ferruginous duck parent

coloured eye and pale flank from pochard parent

This hybrid is fertile and could breed with either species or with another hybrid. This results in individuals with a range of characteristics.

Duck species will occasionally interbreed to produce individuals with a range of characteristics.

The European herring gull and lesser black-backed gull rarely interbreed, even where they nest together. So we say they are different species. But they are the ends of a ring species of gulls that surrounds the North Pole. Neighbouring 'species' of gulls in the ring interbreed frequently, producing a continuous range of characteristics from one end of the ring to the other.

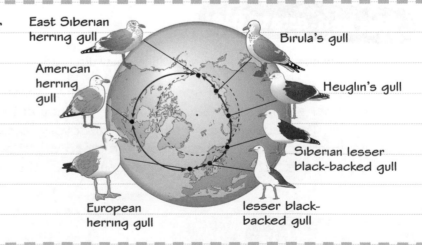

East Siberian herring gull

Birula's gull

American herring gull

Heuglin's gull

Siberian lesser black-backed gull

European herring gull

lesser black-backed gull

Now try this

target D-A

1. Look at the artwork of the hybrid duck above.
   (a) Suggest why, on first sight, this might be classified as a new species. **(2 marks)**
   (b) Give two reasons why it is not classified as a separate species. **(2 marks)**
   (c) Explain why it is difficult to classify some duck species as separate species. **(2 marks)**

Reasons for variety

Different species are adapted to living in different environments. Organisms that live in extreme environments need special adaptations.

Deep-sea hydrothermal vent organisms	
Environment	Adaptations
little light	no eyes
extremely hot water	sense organs that detect dangerously high temperatures
acidic water, at high pressure and full of minerals	soft parts strengthened by iron scales
low oxygen concentration in water	haemoglobin to help take oxygen from water

Polar organisms	
Environment	Adaptations
white snow and ice	white fur for camouflage
very cold in winter	to reduce the amount of heat leaving the body organisms may have: • thick fur • extra fat below skin • bulky body • small ears
slippery ice	wide feet with rough soles to help grip the ice

Types of variation

Discontinuous variation

Characteristics controlled by genes (genetic variation), e.g. blood group, gender.

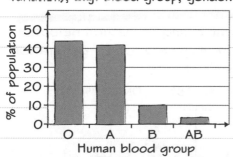

An **acquired characteristic** is one that is changed by the environment.

Continuous variation is where a characteristic varies gradually and continuously between extremes, e.g. height or weight.

A **normal distribution curve** has a bell shape, showing that the most common variation lies between two extremes, with fewer individuals having variations that are near to each extreme.

Worked example

Describe the type of variation shown in this graph, and explain why it shows an acquired characteristic.

The graph shows continuous variation in height between 4 cm and 8 cm. This bell shape is called a normal distribution curve, which shows that the most common height is between 5 and 6 cm, and that the frequency of each height group decreases from the most common group to the extremes of the range. Height is an acquired characteristic because it is affected by differences in the environment.

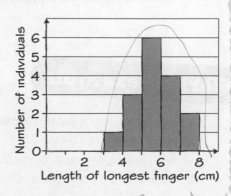

Now try this

target D-B

1. Arctic foxes have thick fur and are white in winter. Explain how these characteristics are adaptations to its environment. **(2 marks)**

2. Write your own definitions for these terms: **(a)** discontinuous variation
(b) acquired characteristic   **(c)** normal distribution curve. **(3 marks)**

Evolution

Natural selection

Individuals of a species show variation. This can mean that some individuals will be better able to survive in their environment and produce more healthy offspring than others. This is natural selection, where the environment (including climate and other organisms) selects which individuals pass on their genes to the next generation.

Being 'better adapted' means being better able to survive, such as surviving the climate, competing more successfully for a limited amount of food, or escaping from predation.

Adults usually produce more young than the environment can support when they are adults (overproduction). This produces a 'struggle for existence' by the young.

Some individuals have inherited advantageous variations in characteristics that are better adapted to the environment. These individuals will have a better chance of survival to adulthood.

Individuals with variations that are not as well adapted to the environment will be less likely to survive.

Individuals with advantageous variations will pass their genes on to their young. The young may inherit the advantageous variations.

These individuals will not produce young.

More individuals will have these advantageous variations in the next generation.

Evolution

Charles Darwin (1809–1882) suggested that if the environment changes, natural selection will result in characteristics of a species changing gradually from generation to generation. This change is called evolution. New evidence supports Darwin's theory.

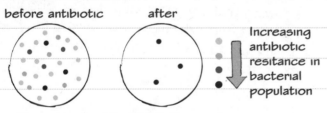

before antibiotic   after   Increasing antibiotic resitance in bacterial population

The antibiotic kills all but the most resistant bacteria. So only these bacteria survive to reproduce.

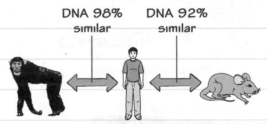

DNA 98% similar   DNA 92% similar

Organisms that are more closely related share a greater proportion of similar DNA.

Worked example

Explain how geographical isolation could result in the evolution of new species.

Some individuals get separated from the rest of the population (e.g. on a different island).

Different conditions in the new place will select for different variations.

Evolution by natural selection will change the characteristics of the individuals in the new place.

Eventually the individuals in the new place may change so much that they could not breed with the rest of the species if they were together. So they would be a new species.

Now try this

target D-B

1. Describe the difference and link between *natural selection* and *evolution*. **(4 marks)**

target C-A

2. Explain how a similarity in DNA between species provides evidence for evolution. **(2 marks)**

Genes

Inside a cell

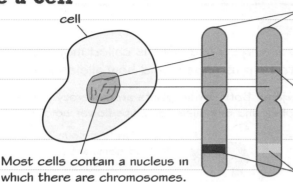

cell

Most cells contain a nucleus in which there are chromosomes.

There are two copies of each chromosome in body cells – each copy has the same genes in the same order along its length (except chromosomes that determine sex).

A gene is a short piece of DNA at a particular point on a chromosome – a gene codes for a characteristic, e.g. eye colour.

A gene may come in different forms, called alleles, that produce different variations of the characteristic, e.g. different eye colours.

Alleles

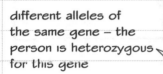

different alleles of the same gene – the person is heterozygous for this gene

chromosomes of the same type are the same size and have the same genes in the same order

these genes have the same allele on both chromosomes – the person is homozygous for these genes

EXAM ALERT!

Be careful not to confuse the meanings of *heterozygous* and *homozygous*.

Students have struggled with this topic in recent exams – **be prepared!**

Result**Plus**

Genetic definitions

The gene for coat colour in rabbits has different alleles. The allele for brown colour (B) is dominant over the allele for black colour (b). The table shows all the possible genotypes and phenotypes for these alleles.

Genotype shows the alleles (forms of the genes) in the individual. Remember that each body cell has two genes for each characteristic – either two alleles that are the same or two that are different.

Genotype	Phenotype
BB	brown coat
Bb	brown coat
bb	black coat

Phenotype means the characteristics that are produced, including what the individual looks like.

The effect of the dominant allele will show when at least one copy is present in the genotype.

The effect of the recessive allele will only show when two copies are present in the genotype.

A dominant allele is not bigger or stronger than a recessive allele. It is dominant because you see its effect on the characteristic even if you only have one copy of it.

Now try this

target
D-B

1. Distinguish between the terms *chromosome*, *gene* and *allele*. **(3 marks)**

2. A pea plant has a recessive allele for white flower colour and a dominant allele for purple flower colour.
   (a) Identify if the plant is homozygous or heterozygous for flower colour, and explain your answer. **(2 marks)**
   (b) State the phenotype of the plant for flower colour. Explain your answer. **(2 marks)**

Explaining inheritance

Monohybrid inheritance

Sometimes a characteristic is controlled by a single gene. This is called monohybrid inheritance. We can use a genetic diagram to help us understand how alleles are inherited.

Body cells contain two alleles for each gene. Both parent plants are heterozygous – they have one allele for purple flower colour and one allele for white flower colour.

parent plants	Purple colour is dominant (R). White colour is recessive (r).
pollen grains egg cells	
different possible gametes	Half the gametes contain one allele. The other half contain the other allele.
possible combinations	
genotype ———	
phenotype ———	

Worked example

Green seed pod (G) is dominant to yellow seed pod (g). Complete the Punnett square to show the possible offspring for plants heterozygous for seed pod colour, and calculate the (a) ratio, (b) probability and (c) percentage of possible offspring genotypes and phenotypes.

		parent genotype Gg	
	parent gametes	G	g
parent genotype Gg	G	GG green	Gg green
	g	Gg green	gg yellow

(a) Genotype 1 GG : 2 Gg : 1gg
   Phenotype 3 green : 1 yellow

(b) Genotype 1/4 (1 out of 4) GG, 2/4 (1/2) Gg, and 1/4 gg
   Phenotype 3/4 green and 1/4 yellow pods

(c) Genotype 25% GG, 50% Gg, 25% gg
   Phenotype 75% green and 25% yellow

A **Punnett square** is a different way of showing the same information about how genotype is inherited and what effect this has on the phenotype.

Genetic diagrams and Punnett squares only show *possible* offspring, not the *actual* offspring from these parents.

EXAM ALERT!

Take great care to complete the square correctly. Make sure you use the right letters.

Students have struggled with questions similar to this – **be prepared!**   ResultsPlus

---

Now try this

1. A heterozygous rabbit with a brown coat was bred with a rabbit with a black coat (homozygous recessive). The four baby rabbits were all black.
   (a) Use a diagram to calculate the predicted outcome of this cross.   **(3 marks)**
   (b) Comment on the difference between this and the actual outcome.   **(2 marks)**

Genetic disorders

Some genes have faulty alleles that cause health problems. These are genetic disorders.

Sickle cell disease

Caused by having two copies of a recessive allele for the haemoglobin gene, which causes red blood cells to become sickle-shaped. People with the disease:
- become short of breath and tire easily
- have painful joints if red blood cells get stuck in capillaries
- have reduced blood flow if red blood cells block a blood vessel, which may cause damage to body tissues, heart attack, stroke or even death.

Cystic fibrosis

Caused by having two copies of a recessive allele for a cell membrane protein. This makes the mucus that lines tubes in the lungs and other parts of the body much thicker and stickier than normal. This can:
- increase the risk of lung infections
- prevent enzymes getting into the digestive system to break down food, which can lead to weight loss.

Family pedigrees

A family pedigree is a diagram that shows the inheritance of a characteristic in a family. Pedigree analysis can help us predict the chance that someone has inherited a particular allele.

Ethan inherited his alleles from Arun and Beth. But they don't have the disease, so they must both be carriers (have one copy of the faulty allele).

Three generations are shown in this pedigree. Arun and Beth are the oldest generation.

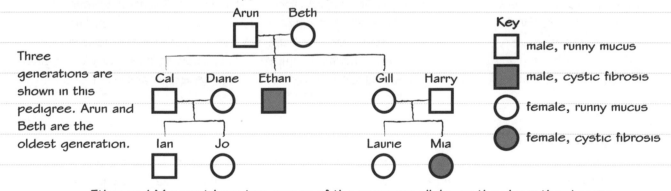

Key
☐ male, runny mucus
■ male, cystic fibrosis
○ female, runny mucus
● female, cystic fibrosis

Ethan and Mia must have two copies of the recessive allele, as they have the disease.

Evaluating risks

We can use pedigrees to evaluate the risk of inheriting a disease. In the pedigree above, Laurie's parents must both be carriers of the CF allele. So Laurie may be a carrier or homozygous for the non-CF allele. If there is no history of CF in her partner's family, the chance that he is a carrier is small. So the risk of any of their children inheriting CF is very small.

Now try this

target D-A

1. People with cystic fibrosis need frequent doses of antibiotics for lung infections. Explain why. **(4 marks)**

target C-A

2. After a pedigree analysis, some people have their DNA tested for the allele they are worried about. Describe the limitations of pedigree analysis compared with DNA testing. **(3 marks)**

Biology extended writing 1

To answer an extended writing question successfully you need to:

☑ Use your scientific knowledge to answer the question

☑ Organise your answer so that it is logical and well ordered

☑ Use full sentences in your writing and make sure that your spelling, punctuation and grammar are correct.

Worked example

Scientists do not always agree on the best way to classify organisms. One thing they *do* agree on is that animals can be described as vertebrates or invertebrates, depending on whether they have a backbone or not. Vertebrates can then be divided up again into different classes.

Describe how scientists have divided vertebrates into different groups. **(6 marks)**

Sample answer 1

Lots of animals are vertebrates. This includes mammals, birds, reptiles and fish. But all these types of vertebrate are different – for example, mammals are warm-blooded, but reptiles are cold-blooded. Also, mammals usually give birth to living babies, but other vertebrates usually lay eggs.

This is a basic answer, although there are some good parts to it. It mentions several types of vertebrate, and includes two ways in which they have been classified. However, it doesn't give very much detail. Also, it does not always use the correct scientific terms, such as homeotherm and poikilotherm. To improve this answer, more could be said about the way in which the organism takes in oxygen, or how its eggs are fertilised.

Sample answer 2

Scientists classify vertebrates by the methods used to reproduce, to get oxygen and to control their body temperature. Most vertebrates reproduce by laying eggs, but most mammals give birth to live young. Vertebrates get their oxygen through either lungs or gills. Vertebrates that live in water – like fish – have gills, but other vertebrates all have lungs. Mammals and birds can both control their own body temperature – they are homeotherms. Other vertebrates, like reptiles and fish are poikilotherms. But there are exceptions. For example, the duck-billed platypus is a mammal but it lays eggs.

This is an excellent answer. It covers the three main ways in which scientists have classified vertebrates. It is also good because it includes an example to show that classification does not always work.

Now try this

1. Explain how the binomial system of classification helps us to classify organisms accurately and how classification helps to conserve the biodiversity of animal life. **(6 marks)**

Biology extended writing 2

Worked example

David and Emily are a couple. They are both heterozygous for cystic fibrosis, a genetic disorder. They decide they would like to start a family and go to their doctor to ask advice.

Discuss how the doctor would use genetics to help David and Emily make decisions about having children.

(6 marks)

Sample answer 1

David and Emily both carry the cystic fibrosis gene – they are both Cc. If they have a child, there is a chance that it would have cystic fibrosis. The doctor might tell them not to have a child if it is going to be ill.

This is a basic answer. It doesn't use correct scientific words properly. A better answer would show a calculation of the probability of having a child with cystic fibrosis. To answer the question fully, more information should be given about the advice the doctor might give.

Both parents carry one cystic fibrosis allele and one 'normal' allele. Their genotype is Cc. The doctor may use a Punnett square to show them the probability of them having a child with CF. In this case, the chance is 25% – if both parents pass on the CF allele to the child. We can see this from the Punnett square.

The doctor will tell them that there is a 1 in 4 chance of them having a child with CF.

The doctor will explain what this means in terms of the risk of having a child with CF. He can help them to decide whether to go ahead with having a child.

This is a very good answer. It shows the probability of the child inheriting CF. It could be improved by drawing a Punnett square which will help explain how the calculation was done.

Now try this

1. Dora was investigating the heights of the people in her class. She measured the height of all 30 people in her class. She recorded the measurements in a tally chart.

Height range (cm)	145–149	150–154	155–159	160–164	165–169	170–174	175–180
Number of people	1	2	7	10	6	3	1

Comment on the pattern in the data and explain how different types of variation may have produced this pattern of data.   **(6 marks)**

Check carefully what the question is asking you to do – in this case, it's to look at types of variation (genetic or environmental) and say how they contribute to the pattern in the data.

13

Homeostasis

Homeostasis maintains some conditions inside the body at a constant level. Negative feedback mechanisms respond to a change in a condition to help bring the condition back to the normal level.

Osmoregulation

Osmoregulation controls how much water is lost in urine, and so controls the amount of water in the body.

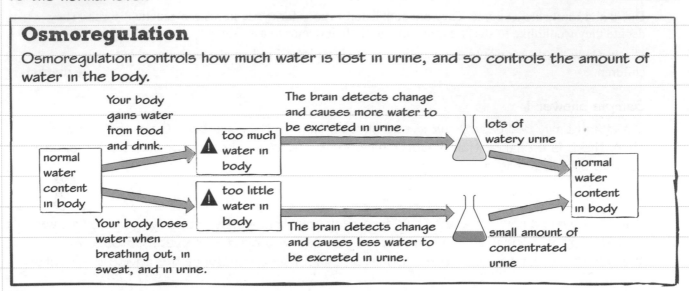

Thermoregulation

Thermoregulation keeps core body temperature steady at around 37 °C. This is important because enzymes in the main organs are most active at this temperature.

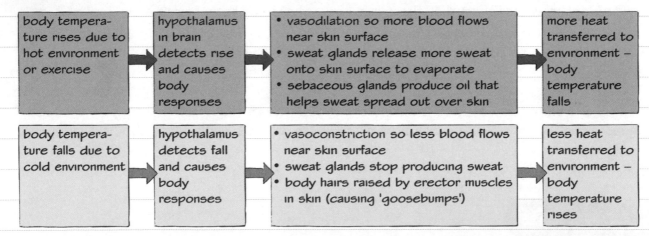

Now try this

target **D-B**

1. Explain why it is important for the enzymes in our bodies that our internal temperature is fairly constant. **(2 marks)**

2. A pale-skinned person may look pink after exercise.

   (a) Identify what causes this change. **(1 mark)**

   (b) Describe and explain what effect this has on body temperature. **(2 marks)**

target **C-A**

3. Describe osmoregulation in terms of a negative feedback system. **(3 marks)**

Sensitivity

The human nervous system includes...

- the central nervous system – brain and spinal cord,

- the sense organs, such as the eyes and ears, which contain receptors that can detect a change in the environment (called a stimulus) and produce an electrical impulse,

- the nerves that join the central nervous system to the sense organs and effector organs – these are made up of bundles of nerve cells, or neurones.

Watch out! Don't confuse the *spine*, which is the bony structure in your back, with the *spinal cord*, which is made of nerves. The spinal cord lies inside the spine for protection.

Types of neurones

There are three main types of neurones. Sensory neurones carry impulses to the central nervous system. Motor neurones carry impulses from the central nervous system to effector organs. Relay neurones are found only in the central nervous system.

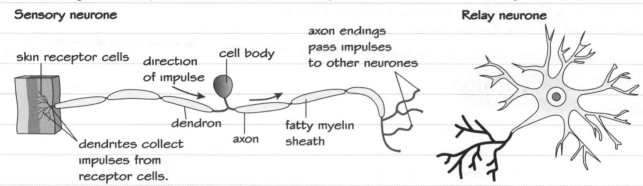

Sensory neurone

skin receptor cells | direction of impulse | cell body | axon endings pass impulses to other neurones

dendron | axon | fatty myelin sheath

dendrites collect impulses from receptor cells.

Relay neurone

Worked example

Explain how the structure of dendrons and axons is related to their function of carrying nerve impulses.

You will get no marks for writing about nerves carrying messages. You should always use the term 'nerve impulses' instead.

The long dendron carries the nerve impulse from receptor cells. The long axon carries the nerve impulse to other neurones. The fatty myelin sheath insulates the neurone, so that the electrical impulse is carried quickly to the end of the axon.

Now try this

target
D-C

1. Compare the roles of sensory, motor and relay neurones in the nervous system. **(3 marks)**

target
C-A

2. Explain how the structure of a sensory neurone is related to its function. **(2 marks)**

Responding to stimuli

Some parts of the skin are more sensitive than others. This can be tested by touching different parts of the skin with two points and finding out how far apart the points have to be so that they can be detected as two points instead of just one.

Synapses

The point where two neurones meet is called a synapse. There is a small gap between the neurones. The electrical nerve impulse cannot cross this gap, and the impulse is carried by neurotransmitters.

(1) Electrical nerve impulse reaches end of axon.

synapse

(2) Electrical impulse causes chemical neurotransmitter to be released into gap between neurones.

(3) Neurotransmitter causes new electrical impulse in next neurone.

The reflex arc

A reflex arc is the simplest neurone pathway from a receptor cell to an effector organ.

Worked example

The diagram shows a reflex arc.

(a) Identify structures A, B and C in the diagram.

(b) Explain how this reflex helps survival.

(a) A: sensory neurone, B: relay neurone, C: motor neurone.

(b) The small number of neurones in the reflex arc makes the response very fast, so you will quickly move your hand away from the flame. This helps survival because it reduces the risk of being burnt.

A — spinal cord
C
B
biceps muscle
heat receptor in the skin

The knee-jerk reflex arc contains only two neurones, but most contain at least three neurones.

EXAM ALERT!

Make sure you follow the command word. For example, here, it asks you to 'explain'. This means that you will often need to use the word 'because' in your answer.

Students have struggled with questions similar to this – **be prepared!**   ResultsPlus

Now try this

target D-B

1. Explain why the nervous system includes chemical neurotransmitters.
   (2 marks)

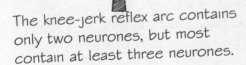
target C-A

2. Explain how a reflex arc allows us to detect a stimulus and then respond.
   (3 marks)

Hormones

Hormones are 'chemical messengers'. They are produced by endocrine glands and released into the blood. They travel around the body in the blood until they reach their target organs. The hormone then causes the target organ to respond, e.g. by releasing another chemical.

Different hormones have different target organs and cause different responses.

Nerves and hormones both help us to respond to changes in the environment and in our bodies. Remember: nerve impulses are electrical signals in neurones, hormones are chemicals carried in the blood.

Blood glucose regulation

Blood glucose regulation is another example of homeostasis. It is controlled by two hormones: insulin and glucagon.

Watch out! **Glycogen** (the carbohydrate) and **glucagon** (the hormone) look similar, but you must use the right word in an answer to get the marks.

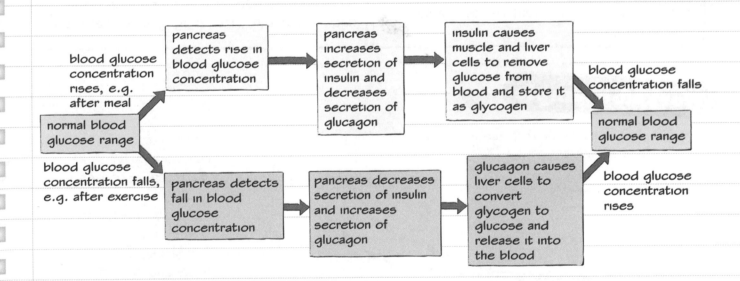

1. Name the endocrine gland that secretes glucagon, and the target organ that the hormone affects. **(2 marks)**

2. For each of the following, suggest two possible reasons why they have happened.

(a) There is a rise in blood glucose concentration. **(2 marks)**

(b) There is a fall in blood glucose concentration. **(2 marks)**

3. The control of blood glucose concentration is an example of negative feedback. Explain why. **(2 marks)**

Diabetes

A person who cannot control their blood glucose concentration properly has a condition known as diabetes. There are two main types of diabetes.

Type 1 diabetes

People with Type 1 diabetes do not produce any insulin in their pancreas and so have to inject insulin into subcutaneous fat (the fat just below the skin). They have to work out the right amount of insulin to inject so that blood glucose concentration is kept within safe limits.

> Glucose is used by cells during exercise, so less insulin is needed then. The more sugar there is in a meal, the more glucose enters the blood and the more insulin is needed.

Type 2 diabetes

People with Type 2 diabetes make insulin, but their liver and muscle cells have become resistant to it (they don't respond to it properly). Most people with Type 2 diabetes control their blood glucose concentration by:

• eating foods that contain less sugar

• exercising.

BMI and Type 2 diabetes

Body mass index (BMI) is calculated using this equation:

$$BMI = \frac{weight\ in\ kilograms}{(height\ in\ metres)^2}$$

People who have a BMI over 30 are said to be obese. Obesity is linked with many health problems, including Type 2 diabetes.

Worked example

Use the graph to evaluate the correlation between obesity and Type 2 diabetes.

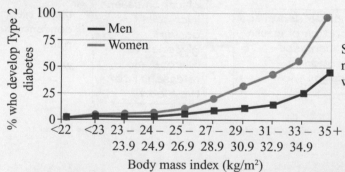

Sample sizes
men: 51 529
women: 114 281

The risk of developing Type 2 diabetes increases as BMI increases. It increases faster for women than for men. The large sample sizes and the smooth curves suggest that this is a strong correlation.

Now try this

 target D–C

1. Tom is 1.8 m tall and weighs 100 kg. Calculate his BMI and say whether or not he is obese. **(3 marks)**

 target D–B

2. Explain why exercise is recommended for people with diabetes. **(3 marks)**

 target B–A*

3. Many health professionals advise that weight control is needed to prevent a huge increase in cases of diabetes over the next decade or two. Evaluate this advice. **(4 marks)**

Plant hormones

Tropisms

A tropism is a plant's response to a stimulus (a change in the environment) by growing. A positive tropism is when the plant grows *towards* the stimulus.

- Plant shoots show positive phototropism because they grow towards light.
- Plant roots show positive gravitropism because they grow downwards – towards the pull of gravity. (Gravitropism is also called geotropism.)

Plant hormones

Plant hormones or plant growth substances are chemicals that cause changes in plants.

- Auxins make cells grow longer.
- Gibberellins can make plant shoots grow longer. They also control when seeds germinate.

EXAM ALERT!

You should describe the effect of light on auxin, and how this affects cell elongation, not just link the direction of light to the curvature of the shoot.

Students have struggled with this topic in recent exams – **be prepared!**   Result$ Plus

Auxins and tropisms

Auxins are affected by light and cause phototropism in shoots. In a shoot, where light is coming from one side:

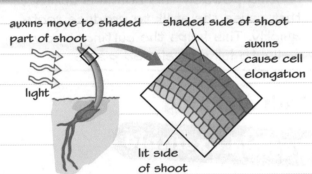

auxin is produced in cells near the top of a shoot

auxins move to shaded part of shoot

shaded side of shoot

auxins cause cell elongation

light

lit side of shoot

Worked example

The graph shows the results of an investigation on the effect of gibberellin on shoot growth. Describe the trends shown by the graph, and comment on how reliable these results are.

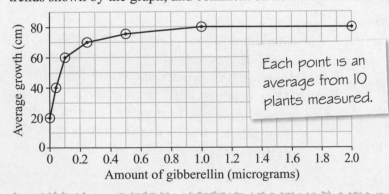

Each point is an average from 10 plants measured.

Growth increases with increasing gibberellin up to a maximum of about 1 microgram. Beyond this, adding more gibberellin has no effect. As each point is an average of 10 plants, this helps to average out random variation in the shoots and makes the results more reliable.

Now try this

target D-C

1. Explain how auxin helps a shoot to grow towards light shining from one side.
(3 marks)

target C-A

2. Explain the advantage of gravitropism to plant roots.
(2 marks)

Uses of plant hormones

We use plant hormones to control the way plants grow.

Selective weedkillers

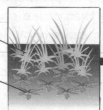

narrow-leaved crop plant

broad-leaved weed plant

selective weedkiller kills broad-leaved plants but not narrow-leaved plants

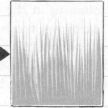

Weed plants compete with crop plants for water and minerals from the soil.

The crop plants get more water and minerals, and so grow better.

Rooting powder

Gardeners take cuttings (small pieces) of plants to grow into new plants. They dip the stalk end of the cutting into rooting powder. This contains auxins that cause the stalk to produce roots quickly. This helps the cuttings grow well into fully developed plants.

Seedless fruit

Hormones sprayed on to flowers can stop seeds developing in the fruits. They can also make the fruit grow larger. Many people prefer large seedless fruit, so they are worth more money.

Fruit ripening

Worked example

The graph shows the effect of a plant hormone sprayed on tomato fruits after picking. Describe what the graph shows, and explain how a hormone like this can be useful to farmers.

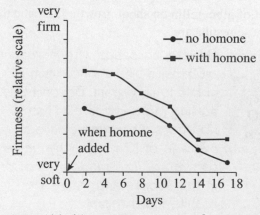

The graph shows that tomato fruits get softer after they are picked, but that those sprayed with hormone stay firmer for longer than those that are not sprayed. Spraying tomatoes will keep them firmer for longer during transport, so that they will be ripe at the right moment for selling.

There are other possible answers. For example, you could say that it stops the fruit getting squashed during transport. What matters here is an answer that shows how the hormone is useful to us. This often links to the fruits being worth more money when they are sold.

Now try this

1. Explain why selective weedkillers are useful.
   (2 marks)

2. Ripe bananas give off a gas called ethene. Pears stored next to ripe bananas ripen more quickly than pears stored next to unripe bananas. Explain why.
   (2 marks)

Biology extended writing 3

Worked example

Some of the foods that we eat contain sugars such as glucose. Our bodies use glucose to make energy. We need energy all the time, even when we are asleep. However, we only need to take in food two or three times a day.

Explain how the body makes sure that the amount of glucose in the blood stream is neither too large nor too small. **(6 marks)**

Sample answer 1

When we eat, sugar gets into our blood. The pancreas makes a chemical called insulin. This breaks up all the sugar again. Diabetic people don't have that chemical, so they need to inject themselves with insulin. We need to eat often so that we keep some sugar in our blood.

This is a basic answer. It gives some information about diabetes, but this isn't what the question asks. You need to make sure that the answer you give is focused on the question. The rest of the answer gives some information about insulin and where it is produced. To improve, the answer should say more about how the body responds to low levels of blood sugar. The answer would also be better if it used correct scientific words such as 'hormone' instead of 'chemical' and 'glucose' instead of 'sugar'.

Sample answer 2

Carbohydrates in our food are broken down to glucose, which is absorbed into the blood. As the concentration of glucose in the blood rises, the pancreas senses the change and releases a hormone called insulin. This hormone travels to the liver in the blood. The liver starts removing glucose from the blood and turning it into glycogen. Glycogen is then stored in the liver. When blood glucose levels get very low, the pancreas senses the change and produces another hormone called glucagon. This hormone also travels to the liver, where glycogen is turned back into glucose and released back into the blood. This is an example of negative feedback.

This is an excellent answer. It is very clear, uses all the correct terminology and gives a thorough description of the process involved in blood glucose regulation.

Now try this

1. **Asma:** 'Humans and animals use hormones like insulin and adrenaline, but plants don't have hormones because they don't have blood.'

   Will: 'Plants do have hormones – they use them to help them grow and develop. Farmers also use plant hormones to help them grow crops.'

   Explain why Will is right by describing how plant hormones help plants grow and how they can be used to help produce crops. **(6 marks)**

Effects of drugs

A drug is a chemical that affects the central nervous system. So a drug can change your psychological behaviour, which is the way you think or feel. For example, narcotics are drugs that make you feel sleepy.

Some drugs cause addiction. This means that the person becomes dependent on taking the drug and feels they cannot function properly without it.

The damage caused by smoking

nicotine is addictive

carbon monoxide reduces how much oxygen the blood can carry

chemicals in tar are carcinogens that cause cancers, particularly of mouth and lungs

Different types of drugs

Type of drug	Example	Effect
painkiller	morphine	block nerve impulses from pain receptors so you don't feel as much pain
stimulant	caffeine	increases neurotransmission at synapses, so increases speed of reaction, which reduces reaction time
depressant	alcohol	slows the activity of the brain, so increases reaction time
hallucinogen	LSD	change perception, e.g. colours seem brighter; can make it difficult to distinguish between real or not real

Watch out! Reaction time *decreases* (gets faster) as speed of reaction *increases*.

The effect of caffeine on reaction time can be measured by timing how long it takes to catch a dropped ruler after drinking normal cola or caffeine-free cola.

Worked example

A study of all the American men who were 40 in 2004 looked at the number of deaths before 2005. It separated the results into smokers and non-smokers. It showed that, compared with non-smokers, 7 times as many smokers died of heart disease, 4 times as many smokers died of lung cancer, but an equal number died after an accident.

Two factors show a correlation when they change in a similar way, e.g. as one increases so does the other.

The data suggest that smoking is correlated (linked) to an increased risk of heart disease and of lung cancer, but not an increased risk of a fatal accident.

As the results come from a large sample of men, they should give a reliable picture.

Now try this

target D-C
1. Describe the effect of caffeine on reaction time. Explain your answer. **(3 marks)**

target D-B
2. Explain why many smokers find it hard to give up smoking. **(2 marks)**

target C-A*
3. Expectant mothers who smoke are more likely to give birth to babies of a lower weight than non-smokers. Explain why this happens. **(4 marks)**

The effects of alcohol

Drinking a lot of alcohol at one time can cause short-term harm including:
- blurred vision
- slow reactions
- lowered inhibitions so you take more risks.

Drinking a lot of alcohol over a long period can cause long-term harm including:
- brain damage
- liver cirrhosis.

Worked example

The graph shows results from a study of the risk of having an accident with different amounts of alcohol in the blood. Describe the trend in the data shown in the graph. Use your scientific knowledge to explain the trend.

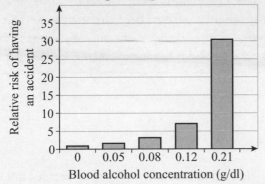

As blood alcohol increases, the risk of being involved in an accident increases. The risk of an accident with 0.21 g/dL of blood alcohol is around 30 times greater than with no alcohol. Alcohol is a depressant. This means that it slows down your reactions, so you take longer to make a decision. This would increase the risk of an accident.

EXAM ALERT!

Your explanation should use your science knowledge about the effect of alcohol on the body and correct science terminology.

Students have struggled with questions similar to this – **be prepared!** | ResultsPlus

Evaluating data

The graph shows a correlation between deaths from cirrhosis and alcohol consumption in some countries.

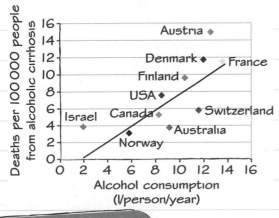

Evaluation of data in graph: the link between cirrhosis and alcohol consumption cannot be confirmed by this graph as other factors might vary between countries (such as food or lifestyle). Also only a few countries are shown – others might change the pattern. So other data are needed to conclude that alcohol consumption causes cirrhosis.

Now try this

target D-B

1. Explain why it is illegal to drive with more than a certain amount of alcohol in your blood.
   (3 marks)

target C-A

2. A large study in the UK concludes that the risk of cirrhosis increases with the amount of alcohol that is drunk. Explain how the graph above makes the conclusion from this study more reliable. **(2 marks)**

Ethics and transplants

Some diseases can be cured by transplanting a healthy organ from a person who has just died into a patient whose organ is damaged. There are never enough healthy organs available for all who need transplants, and many patients will die while waiting for an organ.

Ethics

Ethics is about what you think is right or wrong, fair or unfair. Different people may make a different choice depending on their point of view.

> Ethical argument 1: if two patients need a heart transplant, a patient who is not obese should have priority over an obese patient *because* they are more likely to survive the operation.
>
> Ethical argument 2: everyone should be treated equally *because* an obese patient may not be able to control their weight.

> Ethical argument 1: people should have to say if they <u>don't want</u> their organs to be used for transplants after they die *because* this could increase the number of organs available for transplant.
>
> Ethical argument 2: people should have to say if they <u>do want</u> their organs to be used for transplants after they die *because* the relatives of the dead person may object to the body of their family member being cut up.

Worked example

Many patients who need a liver transplant are alcoholics (addicted to alcohol). A hospital decides not to give liver transplants to patients until they have not drunk alcohol for 6 months.

(a) Discuss a reason for the hospital's decision.

(b) Give one ethical argument against this decision.

(a) The patient may be less likely to drink alcohol after the operation and damage the liver.

(b) An alcoholic finds it difficult to give up drinking alcohol. If their liver is badly damaged, they might die before they are able to control their drinking.

Other answers that show an understanding of the problem will get marks here, e.g. in part (a) you could say that giving up would reduce the addiction to alcohol and this would make the transplant more likely to be successful.

Now try this

target D-B

1. A government is considering a new law to say that organs can be taken from a dead person unless they said while they were alive that they didn't want this.

   (a) Give one argument for and one against this decision. **(2 marks)**

   (b) Explain why this is an ethical decision. **(1 marks)**

target C-A

2. Explain why it is difficult to evaluate an ethical decision. **(2 marks)**

Pathogens and infection

An infectious disease is one that can be passed from one person to another. A pathogen is an organism that causes an infectious disease.

Spreading infection

Pathogens are spread in different ways.

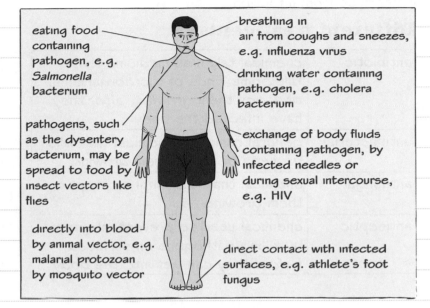

eating food containing pathogen, e.g. *Salmonella* bacterium

pathogens, such as the dysentery bacterium, may be spread to food by insect vectors like flies

directly into blood by animal vector, e.g. malarial protozoan by mosquito vector

breathing in air from coughs and sneezes, e.g. influenza virus

drinking water containing pathogen, e.g. cholera bacterium

exchange of body fluids containing pathogen, by infected needles or during sexual intercourse, e.g. HIV

direct contact with infected surfaces, e.g. athlete's foot fungus

Worked example

Malaria is a disease that affects humans. Identify the organism that causes the disease and explain how it is spread.

Malaria is caused by a protozoan pathogen. It is carried by the *Anopheles* mosquito, which acts as a vector. When the mosquito bites a human, the protozoan can get into the human blood.

Watch out! Vectors are not pathogens. Vectors carry pathogens from one person to another. Vectors are one way that infectious diseases are spread.

Protection against pathogens

Physical barriers	Chemical defences
skin forms a protective barrier	hydrochloric acid in stomach kills pathogens in food and drink
mucus in breathing passages and lungs trap pathogens	
cilia move mucus and trapped pathogens out of lungs	lysozyme enzymes in tears kill pathogens and prevent them entering the eyes

Now try this

target **D-C**

target **D-B**

1. State five different ways in which pathogens can be spread. For each one give an example of a disease which is spread in that way. **(5 marks)**

2. A guide to the outdoors recommends boiling water collected from a stream before drinking. Explain why. **(2 marks)**

3. Explain why a large cut in the skin should be covered with a dressing, but a large bruise does not need covering. **(2 marks)**

Antiseptics and antibiotics

Many plants make chemicals that help to protect them from attack by bacterial pathogens. These chemicals are called antibacterials. We use some of these antibacterials to help protect us from infection.

Different treatments

antibiotic	chemical taken as medicine that kills some kinds of microorganisms, or stops them growing, *after they have infected the body*
antibacterial	general name for chemical that kills bacteria or stops them growing
antifungal	chemical that kills fungi or stops them growing
antiseptic	chemical used to prevent infection by killing pathogens on surfaces *before they can get into the body*

Watch out! Antibiotics cannot be used to treat viral infections such as colds or flu. Viral illnesses can only be treated with antiviral medicines.

Remember: antibiotics treat pathogens inside the body, antiseptics prevent infection by killing pathogens outside the body.

Effect of antibiotics on bacteria

The effect of antiseptics or antibiotics on the growth of microorganisms can be studied in a Petri dish.

antibiotic A

antibiotic B

clear areas where bacteria have been killed by the antibiotic

bacteria growing on agar plate

Antibiotic resistance

Antibiotics kill all but the most resistant bacteria. If these bacteria get into the environment, they can cause antibiotic-resistant infections. The risk of more people getting antibiotic resistant infections can be reduced if:

• antibiotics are only used when essential (e.g. not for viral infections)
• each course of antibiotics is completed, so that the antibiotics and immune system kill all the resistant bacteria.

Misuse of antibiotics has led to the development of bacteria resistant to many antibiotics, such as MRSA.

If you are asked to evaluate evidence for a link between developing resistance and misuse of antibiotics, remember that a good study will have a large sample size, including people with different medical histories, ages, sex and ethnic backgrounds.

Now try this

1. Look at the agar plate illustrated above. Identify which antibiotic was the most effective at killing bacteria and explain your answer. **(2 marks)**

2. Studies show that more patients in hospitals are infected with MRSA than people out in the community. Suggest why this might happen. **(2 marks)**

Interdependence and food webs

Chemical energy is transferred from organism to organism along a food chain or in a food web. All living organisms are interdependent due to their feeding relationships. These relationships are dynamic (always changing), because a change in numbers of organisms in one trophic level (feeding level) will affect other trophic levels.

Energy transfers

energy is transferred along the food chain as
chemical energy in the food that each animal eats

| producer | primary consumer (herbivore) | secondary consumer (carnivore) |

In a question about food webs, it is important to describe effects of interdependency between organisms, not just say that other trophic levels will be affected.

Energy flow

light energy taken in during photosynthesis used to make new plant tissue

heat energy from respiration transferred to surroundings

chemical energy stored in biomass of food

heat energy from respiration transferred to surroundings

chemical energy stored in faeces and urine

chemical energy stored as new plant tissue (biomass) which can be transferred to herbivores in their food

chemical energy stored as new animal tissue (biomass) which can be transferred to carnivores in their food

The biomass of an organism is the mass of its body tissue.

Pyramids of biomass

A pyramid of biomass is a diagram that shows the amount of biomass (usually as g/m^2) at each trophic level of a food chain. The producer level is the bottom bar, and the other bars show the trophic levels in order.

The amount of biomass at each trophic level along the food chain gets smaller. This is because some energy at each level is transferred as heat energy to the environment.

Food chains are usually no more than 4 or 5 trophic levels long because there is not enough biomass in the top level to provide the energy needed by another trophic level.

The food chain for this pyramid is:
lettuce → caterpillar → thrush

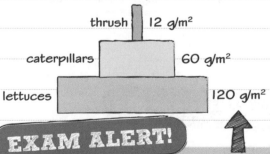

thrush	12 g/m^2
caterpillars	60 g/m^2
lettuces	120 g/m^2

EXAM ALERT!

If you are asked to draw a pyramid of biomass make sure that the sizes of the blocks representing each trophic level are in proportion to the amount of biomass given.

Students have struggled with this topic in recent exams – **be prepared!**   ResultsPlus

Now try this

target D–C

1. (a) Identify two ways that energy is transferred from an animal to the environment. **(2 marks)**
   (b) Identify one way that energy is transferred to a plant from the environment. **(1 mark)**

target C–A*

2. It has been suggested that food chains in tropical rainforests contain more trophic levels than those in temperate areas (like the UK) because rainforests receive more light energy. Explain the reasoning behind this suggestion. **(3 marks)**

Parasites and mutualists

Parasitism

A parasite feeds on another organism (the host) while they are living together. Taking food from the host usually harms it but doesn't kill it. A parasite may be a plant or an animal.

Parasite	Host	Description
flea (animal)	other animals, including humans	fleas feed by sucking the animal's blood after piercing its skin
head louse (animal)	humans	head lice feed by sucking blood after piercing the skin on the head
tapeworm (animal)	other animals, including humans	tapeworms live in the animal's intestine and absorb nutrients from the digested food in the intestine
mistletoe (plant)	trees, e.g. apple	mistletoe grows roots into the tree to absorb water and nutrients from the host

Mutualism

When two organisms live closely together in a way that benefits them both, they are called mutualists.

oxpecker bird benefits by getting food

herbivore benefits from loss of skin parasites

cleaner fish benefit by getting food

the larger fish benefits from loss of dead skin and parasites

nitrogen-fixing bacteria in root nodules are protected from environment and get food from plant

legume plant gets nitrogen compounds for healthy growth from the bacteria

Worked example

Some tube worm species that live near deep sea vents have no mouth, but contain chemosynthetic bacteria. Explain how the relationship between worm and bacteria is mutualistic.

Chemosynthesis is the making of food using the energy from the breakdown of chemicals.

The tube worms benefit by getting food from the bacteria inside them. The bacteria benefit by being protected from the harsh environment around the vent.

Now try this

target
D-C

1. Define the term parasite. **(1 mark)**

2. Explain what is meant by mutualism. **(1 mark)**

target
D-B

3. Explain how the relationship between nitrogen-fixing bacteria and legumes is mutualistic. **(2 marks)**

Pollution

Global human population change

The global human population is increasing. This is because each year the number of babies born is greater than the number of people who die. As human population increases, pollution will increase unless we find ways to reduce the amount of pollutants we produce.

Worked example

The graph shows the best estimate of the number of people in the world since 1950, and three predictions of population size in 2050. Analyse and interpret the graph.

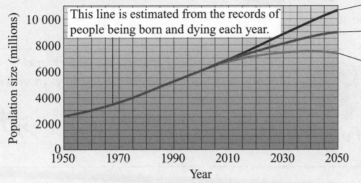

This line is estimated from the records of people being born and dying each year.

Birth rate same as in 2000

Birth rate a little lower than 2000

Birth rate much lower than 2000

All predictions assume death rate stays the same as now.

The graph shows that human population size has risen since 1950 from about 2500 million to nearly 7000 million now. The three different estimates of population size depend on differences in possible birth rates. The high and medium estimates show population size will continue to rise. The low estimate shows that it will level out at about 7500 million in 2030–2040 and then start to fall.

Pollution

Some human activities cause pollution. A pollutant is a substance that can damage the environment and the organisms that live in it. Pollutants include:

• sulfur dioxide gas released from factories and power stations, which pollutes the air

• phosphates and nitrates from sewage and fertiliser, which can pollute water.

Eutrophication

Fertilisers added to fields for crops may get into streams and rivers. This adds phosphates and nitrates to the water, which is called eutrophication.

| Eutrophication causes water plants and algae to grow more quickly. | → | Plants and algae cover the water surface, and block light to deeper water. | → | Deeper plants cannot get light, so they die. | → | Bacteria decompose dying plants and take oxygen from the water. | → | There is not enough oxygen left in the water for fish, so they die. |

Now try this

1. Explain how eutrophication might cause fish to die. **(4 marks)**

2. Discuss the statement 'Over the next 50 years the world will become ever more polluted.' **(3 marks)**

Pollution indicators

Some species are well adapted to living in polluted conditions. Other species can only live where there is no pollution. The presence or absence of these indicator species can show us whether or not there is pollution.

Indicators of air pollution	Indicators of water pollution
Some species of lichen can only grow where there is no pollution. Other species can grow where there is air pollution. So the species of lichen growing on trees can tell you if the air has been polluted.	Bloodworms and sludgeworms can live in water that contains little oxygen. So they are found in polluted water.
Blackspot is a fungus that infects roses. The fungus is damaged by sulfur dioxide in the air. So where there is air pollution, the roses are clear of the fungus.	Stonefly larvae (young stonefly) and freshwater shrimps can only live in water that contains a lot of oxygen. So they are indicators of unpolluted water.

Recycling

Recycling means to re-use materials. Materials that are commonly recycled include metals, paper and plastics.

EXAM ALERT!

Saying that recycling is 'better for the environment' is not enough. You have to link the recycling to saving energy and conserving resources. You are often asked to make links like this.

Students have struggled with exam questions similar to this – **be prepared!**

ResultsPlus

Worked example

Explain how recycling metals, paper and plastics can reduce the amount of landfill and conserve resources.

When metals such as steel and aluminium are recycled, less metal ore has to be extracted from the ground. For most metals, less energy is needed to recycle than produce new metal from its ore, so this also saves resources.

When paper is recycled to make paper or cardboard, this means that trees do not have to be cut down to make new paper. It also takes less energy to recycle paper than make it new from trees.

Recycling plastics saves using oil to make new plastic.

Recycling any material also reduces the amount of waste that we incinerate or dump into landfill tips.

Now try this

1. Explain how each of these species acts as a pollution indicator species:
   (a) stonefly larvae
   (b) blackspot on roses. **(4 marks)**

2. Describe two benefits to the environment of recycling materials. **(2 marks)**

3. Evidence from pollution indicators can be used to locate the source of pollution. Explain what evidence would be needed for this.

(3 marks)

The carbon cycle

A diagram of the carbon cycle shows how the element carbon passes between the environment and living organisms. In the air, the carbon is part of carbon dioxide gas. In organisms, the carbon is part of complex carbon compounds.

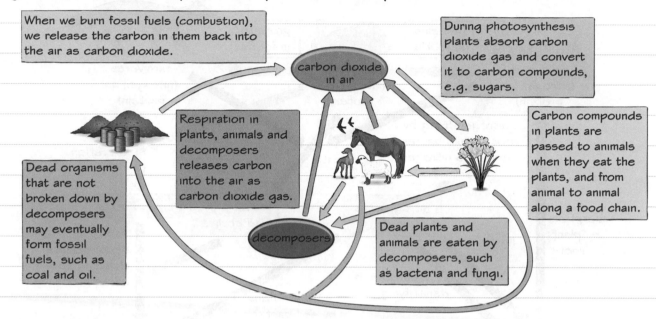

When we burn fossil fuels (combustion), we release the carbon in them back into the air as carbon dioxide.

During photosynthesis plants absorb carbon dioxide gas and convert it to carbon compounds, e.g. sugars.

carbon dioxide in air

Respiration in plants, animals and decomposers releases carbon into the air as carbon dioxide gas.

Carbon compounds in plants are passed to animals when they eat the plants, and from animal to animal along a food chain.

Dead organisms that are not broken down by decomposers may eventually form fossil fuels, such as coal and oil.

decomposers

Dead plants and animals are eaten by decomposers, such as bacteria and fungi.

Worked example

A large forest is cleared by burning. What effects will this have on the amount of carbon dioxide in the air (a) immediately, and (b) over a longer period?

(a) Large amounts of carbon dioxide will be released into the air by the burning (combustion) of the trees.

(b) Less carbon dioxide will be removed from the air than before because the trees would have used some for photosynthesis. So the amount of carbon dioxide in the air is likely to remain high.

EXAM ALERT!

In questions about the carbon cycle, you will be expected to make links between photosynthesis, respiration and combustion, and the amount of carbon dioxide in the air.

Students have struggled with exam questions similar to this – **be prepared!**

Results

Now try this

target
D-C

1. Describe the role of decomposers in the carbon cycle.   **(1 mark)**

target
D-B

2. Explain the effect of respiration, photosynthesis and combustion in the carbon cycle in transferring carbon dioxide to and from the atmosphere.   **(3 marks)**

The nitrogen cycle

A diagram of the nitrogen cycle shows how the element nitrogen moves between living organisms and the environment.

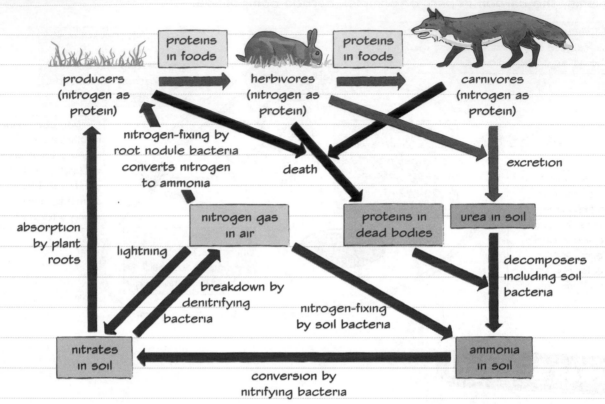

Worked example

Plants grow well in fertile soil. Explain how bacteria help to keep soil fertile.

Plants need nitrogen for making proteins but they can only take in nitrogen in the form of nitrogen compounds such as nitrates. Soil bacteria act as decomposers, releasing ammonia from proteins in dead bodies and from urea. Nitrifying bacteria in the soil turn ammonia into nitrates in the soil. Plants can then use the nitrates to make proteins. Nitrogen fixing bacteria in the soil and root nodules convert nitrogen gas from the air into nitrogen compounds that the plant can use.

EXAM ALERT!

In a question on root nodules, remember that these are related to bacteria that make nitrogen compounds, not the water- or mineral-absorbing abilities of plants.

Students have struggled with this topic in recent exams – **be prepared!** ResultsPlus

Now try this

target E-C

1. Identify two different ways that nitrogen in the air is converted to nitrate in the soil. **(2 marks)**

target C-A

2. Farmers plough crop stubble back into the soil after harvesting. Explain how this helps to improve soil fertility. **(4 marks)**

Biology extended writing 4

Worked example

Although antibiotics have many uses, some doctors feel that overuse of antibiotics can also cause problems. Discuss the benefits and drawbacks of using antibiotics to fight infections. **(6 marks)**

Sample answer 1

Antibiotics are really useful because they stop the germs that cause disease from growing. This means that they can stop you getting the flu and other diseases. However, sometimes the antibiotic doesn't work properly and the germs survive and can still infect you.

This is a basic answer. It misses several keys points. The answer should name the 'germs' that antibiotics kill and it should mention how antibiotic resistance develops. Also, this answer makes the common mistake that antibiotics kill the flu virus — antibiotics do not work on viruses.

Sample answer 2

The benefits of antibiotic drugs are that they either kill bacteria or they prevent them from reproducing. This helps to prevent or fight bacterial infections. The drawbacks are that some doctors give patients antibiotics when they do not have a bacterial disease and that some patients do not finish the antibiotics they are given. Not all the bacteria are killed by the antibiotic and the ones left are generally the ones which are least affected by the antibiotic. This characteristic is then passed to future generations through natural selection. If bacteria which are resistant to antibiotics infect other people then it can be very hard to get rid of the infection. Antibiotic resistance is a big problem in hospitals, because patients can get something like MRSA.

This is an excellent answer. It covers all the key points, and gives a good outline of how antibiotic resistance develops. It also uses correct scientific words such as 'bacteria' and 'resistant'. It could be improved by saying a little more about MRSA.

Now try this

1. Describe the ways in which different pathogens can be passed from one person to another. Your answer should include some named examples. **(6 marks)**

Think about how you would like to present the answer to this question. You might find it easier to draw up a table showing a disease, the type of pathogen which causes it, and the way in which it is spread.

Biology extended writing 5

Worked example

Sean has been suffering from heart disease for several years. He is clinically obese. A doctor at the hospital talks to Sean about the possibility of having a transplant.

Explain the medical and ethical factors that the doctor will consider when deciding whether to recommend Sean for a transplant. **(6 marks)**

Sample answer 1

The doctor will tell Sean that hearts do not come up often. If Sean is very fat, then the doctor may not be happy with giving a new heart to Sean. There could be other people who deserve it more. The doctor wouldn't want to give Sean a new heart if the new heart also got damaged. The doctor might tell Sean to lose some weight.

This is a good answer. It mentions several factors – although in each case it could give more detail. It needs to explain what sort of person might 'deserve it more'. Also, how would Sean losing weight help? The answer could also be split into two parts, talking about medical factors and ethical factors. This helps to make the answer clearer, and also helps you to remember to include both kinds of factor in your answer.

Sample answer 2

Medical issues
The doctor will look at lots of different medical issues. He'll think about how long Sean has been ill and whether he is in great danger of dying. There may be other people who are more ill. Donor hearts don't come up often. Donor hearts don't come up often, so the hospital has to choose who to give the heart to. One factor the doctor will consider is whether Sean is healthy enough to have the operation.

Ethical issues
Ethically, the doctor will need to consider whether Sean is the right recipient for a heart. Sean's weight may be due to a medical condition, so it's not his fault. If Sean can show that he is keen to be more healthy by losing weight, then the doctor may also be more likely to recommend Sean for a heart transplant.

This is an excellent answer. There is a very thorough list of medical factors and the section on ethical issues gives a clear summary of the factors that would be considered. There is a good balance between the two sides of the answer.

Now try this

1. Fizzy cola drinks contain large amounts of caffeine. A group of students wants to investigate the effect that drinking fizzy cola has on reaction time.

   State a hypothesis that the students could make and describe an investigation to test this hypothesis.

   (6 marks)

The early atmosphere

The Earth's first atmosphere

The Earth's first atmosphere was formed by gases produced from volcanoes. It contained large amounts of carbon dioxide and water vapour, hardly any oxygen and small amounts of other gases.

The Earth was very hot to start with, and there were no oceans. As the Earth cooled down the water vapour in the atmosphere condensed to form liquid water. This liquid water became the oceans.

Worked example

Explain why scientists cannot be certain about the composition of the early atmosphere.

There were no humans around to measure the atmosphere, so they have to use clues. There are different sources of information, but not all the evidence leads to the same conclusion. This means that it is difficult to know exactly what the early atmosphere was like and how it has changed.

The evidence includes:

• the gases produced by volcanoes today, which tell us what gases would have been produced by volcanoes in the past

• the atmospheres of other planets and moons in the solar system, where the atmosphere has not been changed by living things

• iron compounds found in very old rocks that could only form if there was no oxygen.

You don't need to remember these different sources of information. You *do* need to remember that there *are* different sources of information, some of which tell us different things.

Adding oxygen and removing carbon dioxide

The atmosphere has changed since the Earth was formed.

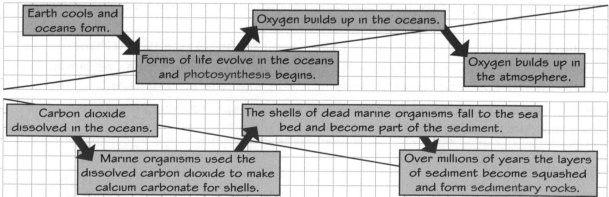

Earth cools and oceans form. → Forms of life evolve in the oceans and photosynthesis begins. → Oxygen builds up in the oceans. → Oxygen builds up in the atmosphere.

Carbon dioxide dissolved in the oceans. → Marine organisms used the dissolved carbon dioxide to make calcium carbonate for shells. → The shells of dead marine organisms fall to the sea bed and become part of the sediment. → Over millions of years the layers of sediment become squashed and form sedimentary rocks.

Now try this

target D-B

1. Explain why scientists think that the Earth's early atmosphere contained a lot of carbon dioxide and no oxygen. **(3 marks)**

2. Explain why the levels of water vapour, carbon dioxide and oxygen did not start to change for many millions of years after the Earth was formed. **(3 marks)**

A changing atmosphere

The amounts of different gases in the atmosphere today are shown in the pie chart. The atmosphere also contains water vapour, but this is not usually included because the amount changes depending on the weather.

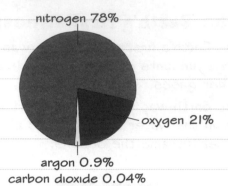

nitrogen 78%

oxygen 21%

argon 0.9%
carbon dioxide 0.04%
traces of other gases

Worked example

There is a lot of limestone in the Earth's crust. Explain how the amount of carbon dioxide in the atmosphere has changed since the Earth was formed, and how this is linked to the formation of limestone.

The atmosphere was almost all carbon dioxide when the Earth was formed, and it has decreased to approximately 0.04%. Some carbon dioxide in the atmosphere dissolved in the oceans. Some marine organisms use dissolved carbon dioxide to make calcium carbonate for their shells. When they die, these marine organisms sink to the sea bed and their shells eventually form limestone. The carbon locked up in the limestone originally came from the atmosphere.

You need to explain the link between limestone and carbon by saying that limestone is made from shells that are mostly calcium carbonate.

Changes in the atmosphere

The composition of the atmosphere today is not constant. It can be changed by human activities.

- Burning fossil fuels releases carbon dioxide into the atmosphere. Sulfur dioxide can also be released from burning coal.
- Farming affects the atmosphere, as cattle and rice fields release methane.
- Deforestation adds carbon dioxide to the atmosphere if the trees are burnt. There are also fewer trees to remove carbon dioxide from the atmosphere.

There are also natural causes of changes in the atmosphere. Volcanoes emit sulfur dioxide and carbon dioxide when they erupt.

Now try this

target
D-B

1. Describe and explain the changes in the proportions of water vapour, carbon dioxide and oxygen in the atmosphere since the Earth was formed.   **(3 marks)**

2. Explain how the following activities can change the atmosphere:
   (a) cutting down forests to provide grazing for cattle.   **(2 marks)**
   (b) generating electricity using fossil fuels.   **(1 mark)**

There are two marks for part (a) of this question, so you need to mention two different changes.

Rocks and their formation

Rocks make up the Earth's crust.

Igneous rocks

Igneous rocks:

- are formed when magma or lava solidifies
- are made of interlocking crystals, which makes them hard and resistant to erosion
- have small crystals if the liquid rock cooled quickly (e.g. basalt)
- have large crystals if the liquid rock cooled slowly (e.g. granite).

Watch out! Magma is liquid rock *beneath* the Earth's surface. Liquid rock is called lava when it is *on* the surface of the Earth.

If the rock cools down *slowly*, the crystals have *more time* to grow bigger.

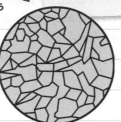

Magnified view of interlocking crystals

Sedimentary rocks

Sedimentary rocks:

- are formed when layers of sediment are compacted (squashed together) over a very long time by being buried under more layers
- erode (wear away) more easily than igneous and metamorphic rocks
- are made of rounded grains and may contain fossils (the remains or traces of living organisms)
- include chalk and limestone, which are natural forms of calcium carbonate.

Magnified view of grains in a sedimentary rock

Metamorphic rocks

Worked example

Explain how metamorphic rocks are formed, and give an example of a metamorphic rock.

Metamorphic rocks are formed from existing rocks by the action of heat (from being buried or from nearby magma) and/or pressure (from being buried), which causes new interlocking crystals to grow. Marble is a metamorphic rock which is formed from chalk or limestone.

Marble is also a natural form of **calcium carbonate**.

EXAM ALERT!

You need to revise how different rocks are formed. For example, to answer this question fully you would need to explain that metamorphic rocks are formed by heat and pressure.

Students have struggled with exam questions similar to this – **be prepared!**   ResultsPlus

Now try this

target D–B

1. Compare how easily igneous and sedimentary rocks are eroded, and use their structures to explain any differences. **(3 marks)**

2. Explain why fossils are found in sedimentary rocks but never in igneous rocks. **(3 marks)**

Limestone and its uses

Using limestone

Limestone is used for making buildings, and as the base for roads and railways. Limestone is mostly made of calcium carbonate ($CaCO_3$), and this substance is an important raw material used to make glass, cement and concrete.

Limestone is heated when making cement and glass. The heat decomposes (breaks down) the calcium carbonate in it to form calcium oxide (CaO) and carbon dioxide (CO_2). This process is called thermal decomposition.

calcium carbonate → calcium oxide + carbon dioxide

Worked example

What is the most important use of limestone?

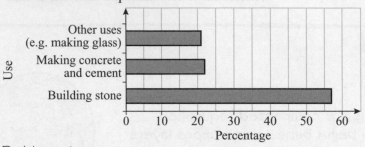

Building stone

Quarrying – good or bad?

We need to quarry a lot of limestone, but quarrying has economic, environmental and social effects. We need to balance the effects with the demand.

Advantages	Disadvantages
✓ limestone is an important raw material	✗ quarries produce dust and noise
✓ quarries provide jobs in the countryside	✗ digging a quarry destroys habitats for wildlife
✓ some limestone is exported to other countries, which helps the UK economy	✗ quarries spoil the scenery (although they are usually landscaped after quarrying has finished and can become wildlife reserves or recreation areas)
✓ jobs in quarries also help local economies, because the workers spend their money in local shops and businesses	✗ lorries taking the limestone away cause extra traffic, noise and pollution

Now try this

target D-B

1. (a) List three uses of limestone. **(3 marks)**
   (b) Which of these uses does **not** require the thermal decomposition of the limestone? **(1 mark)**

2. A quarry company has applied to the local council for permission to dig a new quarry. List two economic factors and two environmental factors that the local council should consider when reaching its decision, and state whether each factor is an advantage or disadvantage. **(4 marks)**

target C-A

3. Write a balanced equation to show the thermal decomposition of calcium carbonate. **(3 marks)**

Formulae and equations

An element is a substance that cannot be split into simpler substances. An atom is the smallest part of an element that can take part in chemical reactions. A molecule is two or more atoms chemically joined together.

A compound is a substance that consists of atoms of different elements chemically joined together. A mixture contains different elements or compounds, but they are not chemically joined.

During a chemical reaction the atoms in the reactants are rearranged to make different products.

EXAM ALERT!

It is important to learn these definitions and other key scientific words. You need to understand the meanings of words such as atom, molecule, element, mixture and compound.

Students have struggled with this topic in recent exams – **be prepared!**

ResultsPlus

Symbols and formulae

Each element has its own chemical symbol. The formula of a compound shows the symbols of the elements it contains and the numbers of their atoms in the molecules. The atoms of some elements form molecules, so these elements have a formula as well.

Remember that gases such as hydrogen, oxygen, chlorine and nitrogen have two atoms in each molecule, so they should be written as H_2, O_2, Cl_2 and N_2.

2H is not the same as H_2. 2H means two separate hydrogen atoms. H_2 means two hydrogen atoms joined into one molecule.

Worked example

In zinc carbonate, there is 1 carbon atom and 3 oxygen atoms for every zinc atom. Write the formula for zinc carbonate.

$ZnCO_3$

Equations for reactions

Chemical reactions can be represented using words or symbols.

hydrogen + chlorine → hydrogen chloride

$$H_2(g) + Cl_2(g) \rightarrow 2HCl(g)$$

State symbols show whether the substances are solids (s), liquids (l), gases (g) or in solution (aq).

There must always be the same number of atoms of each type on each side of the equation.

Now try this

target C-B

1. Common salt (NaCl) dissolves in water.
   (a) Explain whether the salty water is a mixture or a compound. **(2 marks)**
   (b) Write the formula to represent sodium chloride solution. **(1 mark)**

2. Choose two words from this list to describe chlorine gas and explain your answer: atoms, molecules, element, compound. **(3 marks)**

Chemical reactions

Rearranging atoms

In a chemical reaction atoms are rearranged to form different substances. The atoms are not created or destroyed. There are the same number of each type of atom present before and after the reaction — they are just joined up in different combinations. The products of a reaction have different properties than the reactants. As no atoms have been created or destroyed, the total mass of all the atoms involved stays the same.

Worked example

Select the correct balanced equation to show the reaction between hydrogen and oxygen to form water. Tick one box.

☐ $H + O \rightarrow H_2O$

☐ $H_2 + O_2 \rightarrow H_2O_2$

☒ $2H_2 + O_2 \rightarrow 2H_2O$

☐ $H_2 + O_2 \rightarrow 2H_2O$

Hydrogen and oxygen must be written as H_2 and O_2, so this is wrong.

Do not change the formula of a substance to try to balance an equation. Water is always H_2O!

Thermal decomposition

Metal carbonates decompose when they are heated to form an oxide and carbon dioxide. Some carbonates decompose more easily than others (decomposition happens at a lower temperature).

calcium carbonate — hardest to decompose

zinc carbonate

copper carbonate — easiest to decompose

Precipitation reactions

A precipitation reaction happens when two soluble substances react together to form a product that is insoluble. The insoluble product is the precipitate.

lead nitrate + potassium iodide → potassium nitrate + lead iodide
(soluble)   (soluble)   (soluble)   (insoluble – forms a precipitate)

You can use a precipitation reaction in a sealed container to demonstrate that the total mass before and after a reaction does not change.

Now try this

target D-C

1. Write a word equation to show the thermal decomposition of zinc carbonate. **(1 mark)**

target B-A*

2. Magnesium reacts with hydrochloric acid (HCl) to produce a solution of magnesium chloride ($MgCl_2$) and hydrogen gas. Write a balanced equation to show this reaction, including state symbols. **(4 marks)**

3. Write a balanced equation to show the precipitation reaction between lead nitrate and potassium iodide solutions shown in the word equation above. The formulae you need are $Pb(NO_3)_2$, KI, KNO_3 and PbI_2. Include state symbols. **(3 marks)**

Reactions of calcium compounds

This reaction shows the relationships between different calcium compounds.

Limewater is used to test for carbon dioxide. If carbon dioxide is bubbled through limewater it forms insoluble calcium carbonate.
This turns the limewater cloudy.

add CO_2

calcium carbonate — heat

calcium hydroxide

calcium oxide

add water

and CO_2

When calcium carbonate is heated it undergoes thermal decomposition to form calcium oxide and carbon dioxide.

When water is added to calcium oxide a lot of heat is released. The mixture fizzes, steam is given off and the calcium oxide crumbles to a white powder of calcium hydroxide. If more water is added to the reactants, the calcium hydroxide dissolves to form a solution called limewater.

EXAM ALERT!

An exam question may ask you to describe what you see when a reaction happens. So write about what you can *see* – don't write 'heat is given off'.

Students have struggled with this topic in recent exams – **be prepared!**

Results Plus

You need to know these word equations:
calcium carbonate → calcium oxide + carbon dioxide
calcium oxide + water → calcium hydroxide
calcium hydroxide + carbon dioxide → calcium carbonate

Using calcium compounds

Calcium carbonate, calcium oxide and calcium hydroxide can all be used to neutralise acidic compounds. (There is more about neutralisation on page 44.)

Some soils are acidic. Farmers use calcium compounds to neutralise the acidity so their crops will grow better.

This paragraph explains how calcium carbonate is used to clean up the gases. You will learn more about acid rain on page 57.

Worked example

Explain why the gases emitted by coal-fired power stations need to be cleaned up, and how calcium carbonate can be used to do this.

The waste gases from coal-fired power stations contain sulfur dioxide and nitrogen oxides, which are acidic gases. These gases cause acid rain if they get into the atmosphere.

Wet, powdered calcium carbonate is sprayed into the waste gases to neutralise them.

Now try this

1. (a) Describe what you see when carbon dioxide is bubbled through limewater. **(1 mark)**
   (b) Write a balanced equation to show what causes this change. Include state symbols.
   (3 marks)

2. (a) Describe what you would see when a few drops of water are added to a lump of calcium oxide. **(3 marks)**
   (b) Write a balanced equation to show this reaction. Include state symbols. **(3 marks)**

Chemistry extended writing 1

To answer an extended writing question successfully you need to:
- ✓ Use your scientific knowledge to answer the question
- ✓ Organise your answer so that it is logical and well ordered
- ✓ Use full sentences in your writing and make sure that your spelling, punctuation and grammar are correct.

Worked example

The apparatus shown is used to find the percentage of oxygen in the atmosphere. Explain how it works, give an equation for the reaction that takes place and an expected result.

(6 marks)

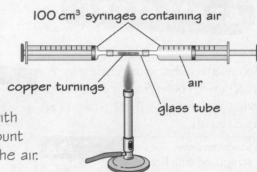
100 cm³ syringes containing air
copper turnings
air
glass tube

Sample answer 1

The air is pushed across the copper and it reacts with the copper. The volume of gas gets smaller. The amount left in the syringes is the percentage of oxygen in the air.

copper + oxygen → copper oxide

This is a basic answer. The first part is correct, but does not give much detail. The last sentence is incorrect, as the oxygen reacts with the copper so that the *change* in the volume of gas represents the proportion of oxygen in the atmosphere. The question asked for a balanced equation, not a word equation. Make sure you read the question carefully.

Sample answer 2

Oxygen in the air will react with the copper when the copper is heated, to form black copper oxide. The other gases will not react.

$$2Cu + O_2 \rightarrow 2CuO$$

The syringes are pushed in and out to pass the air over the copper until the volume stops changing. This shows that all the oxygen in the air has reacted with the copper.

The change in the volume of gas is the volume of oxygen that has reacted. This should be 21 cm³, or 21%.

This is a good answer. It explains the process, includes a correctly balanced equation, and explains how you would know that all the oxygen in the air had reacted.

It could be improved by explaining how to make sure the results are accurate.

Now try this

1. Rocks can be igneous, sedimentary or metamorphic. Describe the processes that can change each type of rock into any of the others, and explain the characteristics of the different rocks formed.

(6 marks)

Indigestion

Stomach acid

Our stomachs produce hydrochloric acid. The acid is used to:

Stomach acid is always hydrochloric acid.

- kill bacteria in our food

- help digestion, as some of the enzymes that break down our food into simpler substances only work in acidic conditions.

Sometimes our stomachs produce too much acid. This causes pain that we call indigestion.

Neutralisation

Indigestion remedies contain antacids that neutralise excess stomach acid. Neutral liquids are neither acid nor alkaline.

Antacids contain substances called bases. Bases react with acids and neutralise them. Some bases are soluble. A base dissolved in water is called an alkali.

acid + base → salt + water

Worked example

An antacid contains magnesium hydroxide. Write the word equation for the reaction between hydrochloric acid and magnesium hydroxide.

hydrochloric acid + magnesium hydroxide
→ magnesium chloride + water

It helps if you remember the general equation for the reactions between acids and bases. Just write out the acid and base that you are given, then work out the name of the salt. Don't forget to include the water!

Testing indigestion remedies

Neutral liquids have a pH of 7. Acids have a pH less than 7. When an indigestion remedy is added to an acid, it will neutralise some of the acid and the pH will increase.

To make this investigation a fair test, you must use the same volume of acid each time at the same concentration, and use one dose of each indigestion remedy. The remedy that produces the highest pH at the end has neutralised the most acid.

Now try this

1. Explain how indigestion remedies cure indigestion. **(2 marks)**

2. Write a balanced equation to show the reaction between magnesium hydroxide ($Mg(OH)_2$) and hydrochloric acid (HCl). The formula for magnesium chloride is $MgCl_2$. Include state symbols. **(3 marks)**

Neutralisation

There are three different types of compound that can be used to neutralise acids:

- metal oxides (such as copper oxide, CuO)
- metal hydroxides (such as sodium hydroxide, NaOH)
- metal carbonates (such as copper carbonate, $CuCO_3$).

> Make sure that you know what scientific words like 'neutralise' mean and then use them correctly in the exam.

Word equations

You need to learn these general equations. You might have to apply them in your exam.

acid + metal oxide → salt + water

acid + metal hydroxide → salt + water

acid + metal carbonate → salt + water + carbon dioxide

> Watch out! These are the *only* kinds of metal compounds that can be used to neutralise acids. And don't forget that carbonates produce carbon dioxide as well as salt and water.

Naming salts

A salt is a compound made from a metal and a non-metal. The non-metal part of the salt comes from the acid. The metal part of the salt comes from the base or alkali used to neutralise it.

- hydrochloric acid produces chloride salts
- nitric acid produces nitrate salts
- sulfuric acid produces sulfate salts

Worked example

Complete these equations:

nitric acid + sodium hydroxide → <u>sodium nitrate</u> + <u>water</u>

$HNO_3(aq) + NaOH(aq) →$ <u>$NaNO_3(aq)$</u> + <u>$H_2O(l)$</u>

> The first part of the name of the base is a metal. This metal is the first part of the name of the salt formed in the reaction.

Hazard symbols

Acids are hazardous substances and can be dangerous if not used properly. Containers of acids have hazard symbols on them. These standard symbols warn people about the hazards. People can then look up the precautions they should take to stay safe when working with these substances.

Now try this

target **D-B**

1. Write word equations for the following reactions:
   (a) copper oxide and sulfuric acid **(2 marks)**
   (b) hydrochloric acid and copper carbonate. **(2 marks)**

target **B-A***

2. Write balanced equations for the reactions in question 1. Include state symbols.
   You will need these formulae: CuO, H_2SO_4, $CuSO_4$, HCl, $CuCO_3$, $CuCl_2$. **(7 marks)**

The importance of chlorine

Compounds can be decomposed (broken up) using electricity. This process is called electrolysis. Electrolysis only works with a direct current (d.c.) such as the current from batteries.

Electrolysis

The apparatus in the diagram can be used to decompose dilute hydrochloric acid. Hydrogen gas is produced at one electrode, and chlorine gas is produced at the other.

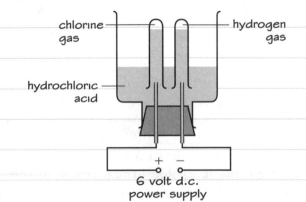

chlorine gas — hydrogen gas

hydrochloric acid

+ −

6 volt d.c. power supply

Using chlorine

Sea water contains a lot of dissolved sodium chloride. Chlorine gas can be obtained from sea water by electrolysis. Chlorine is used to make bleach, and to make plastics such as poly(chloroethene) (also called PVC).

You need to know the chemical symbols and formulae for the substances you learn about. The symbol for chlorine is Cl. This is always a capital 'C' with a small 'l'.

Chlorine gas is made of molecules with two chlorine atoms joined together. The formula for chlorine gas is Cl_2.

Dangers of chlorine

Chlorine is a toxic gas. This makes it useful as a disinfectant because it kills microorganisms. This also makes chlorine hazardous. Chlorine is very useful in industry, and millions of tonnes of it are produced every year. Accidental leaks of chlorine from factories or from tankers could harm or kill many people.

Worked example

Containers of chlorine are labelled with the 'toxic' hazard symbol.

Explain the precautions you should take if you are carrying out electrolysis of hydrochloric acid in the school laboratory.

Toxic

The laboratory must be well ventilated, by keeping the doors and some windows open. Take care not to sniff the gases produced.

These precautions are needed because the electrolysis of hydrochloric acid produces chlorine. Chlorine is a toxic gas, so it will harm you if you breathe it.

The question asks you to 'explain', so you must say *why* the precautions are needed.

Now try this

target C-B

target B-A*

1. Explain why the laboratory must be well ventilated if you are carrying out electrolysis of sea water. **(3 marks)**

2. Write a word equation to show the decomposition by electrolysis of hydrochloric acid. **(1 mark)**

3. Write a balanced equation to show the decomposition of hydrochloric acid. Include state symbols. **(3 marks)**

The electrolysis of water

Electrolysis

Water is a compound. The electrolysis of water produces hydrogen gas and oxygen gas.

Water molecules contain two atoms of hydrogen chemically joined to one atom of oxygen. The formula of water is H_2O. Electrolysis decomposes water into hydrogen (H_2) and oxygen (O_2).

Hydrogen

Hydrogen is a flammable gas. It can explode when it is mixed with air. Containers of hydrogen have the flammable hazard symbol on them.

Flammable

Remember that electrolysis only works with a direct current (d.c.). Alternating current (like the current from the mains supply) does not work.

Testing gases

The gases produced in reactions can be tested to find out what they are.

Gas	Hydrogen	Oxygen	Chlorine
Test	hold a lighted splint in the mouth of the test tube	hold a glowing splint in the mouth of the test tube (light the splint, then blow it out so that the end is just glowing)	hold a piece of damp blue litmus paper in the mouth of the test tube
Result	if the gas is hydrogen, it will explode with a squeaky 'pop'	if the gas is oxygen the splint will relight (burst into flame) again	if the gas is chlorine the paper will turn red and then turn white

Worked example

A student tests a gas by holding a piece of damp blue litmus paper in it for a few seconds. The paper turns red. Student X says this proves the gas is chlorine. Student Y says the gas may be chlorine. Explain who is correct.

Student Y is correct. Chlorine is not the only gas that turns blue litmus paper red.

EXAM ALERT!

If an exam question asks how you would test something always describe the results of a test for a gas, as well as describing how to do the test.

Students have struggled with this topic in recent exams – **be prepared!** ResultPlus

Many gases will turn damp blue litmus paper red. However, it is *only* chlorine that then bleaches it so that it turns white.

Now try this

 target C-B

1. Write a word equation to show the decomposition by electrolysis of water. **(1 mark)**

 target B-A*

2. Write a word equation to show what happens when a lighted splint is put into a test tube containing a mixture of hydrogen and air. **(1 mark)**

3. Write a balanced equation to show the decomposition of water. Include state symbols. **(3 marks)**

Ores

Some metals are found as elements in the Earth's crust. This means the metal is not combined with other elements. Metals found as elements are very unreactive metals such as gold. This means lumps of gold can sometimes be found.

Extracting metals

Most metals are found as part of a compound. For example, iron is found as iron oxide. The metal compounds are mixed up with other compounds in rocks. If a rock has enough of the metal compound to make it worthwhile to extract it, the rock is called an ore.

Metals are extracted from the compounds found in ores. The method used to extract a metal depends on how reactive the metal is.

Iron and aluminium

Metals such as iron are extracted from their compounds by heating the ores with carbon.

iron oxide + carbon
$\rightarrow$ iron + carbon dioxide

More reactive metals such as aluminium are extracted by melting the ore and then carrying out electrolysis (see page 46). Electrolysis decomposes aluminium oxide.

aluminium oxide $\rightarrow$ aluminium + oxygen

Reactivity series

You don't need to learn the order of the elements in the table. You DO need to remember how iron and aluminium are extracted, and that the method depends on the reactivity of the metal. You should also be able to comment on how the way a metal is extracted is linked to the cost of the metal

most reactive
(hardest and most expensive to extract)

least reactive
(easiest and cheapest to extract)

potassium	
sodium	electrolysis of a molten compound
calcium	
magnesium	
aluminium	
zinc	
iron	
tin	heat an ore with carbon
lead	
copper	
silver	
gold	found as the uncombined element
platinum	

Worked example

A student mixes powdered charcoal (a form of carbon) with some copper oxide. After heating the mixture she finds tiny pieces of copper metal.

Explain which of these metals cannot be extracted from their compounds using the same method:

tin　aluminium　magnesium　lead

Aluminium and magnesium cannot be extracted using carbon, because they are too reactive.

Now try this

1. Explain why platinum is not found in the Earth's crust as an ore. **(2 marks)**
2. Cassiterite is an ore containing tin oxide.
   (a) State how the tin is extracted from cassiterite. **(1 mark)**
   (b) Write a word equation to show the reaction. **(1 mark)**

3. Write a balanced equation including state symbols to show the decomposition by electrolysis of aluminium oxide (Al_2O_3). **(3 marks)**

Use the reactivity series given above to help you with these questions.

Oxidation and reduction

Oxidation

Most metals react with oxygen in the air. Sometimes water also takes part in the reaction. This causes corrosion. For example, iron corrodes to form iron oxide.

iron + oxygen → iron oxide

This is an example of an oxidation reaction. The iron has been oxidised, because it has gained oxygen.

Watch out! Many metals corrode. When iron corrodes we call it rusting. Rusting only applies to iron. It is not correct to say that copper or lead 'go rusty'.

Reduction

Many metal ores contain oxides of the metal (see page 47). The metal is extracted from its compound by removing the oxygen.

aluminium oxide → aluminium + oxygen

This is an example of a reduction reaction. The aluminium oxide has been reduced, because oxygen has been removed from it.

Remember: reduction is removing oxygen.

Worked example

Explain how oxidation and reduction can happen in the same reaction. Use the word equation for the extraction of iron as an example.

You need to learn the word equations for the extraction of iron and the extraction of aluminium.

The word equation for the extraction of iron from iron oxide is:

iron oxide + carbon → iron + carbon dioxide

The iron oxide is reduced in this reaction, because oxygen is removed from it. The carbon is oxidised in the reaction.

Reactivity and corrosion

This diagram explains the relationship between the reactivity of a metal and how resistant it is to corrosion.

most reactive										least reactive		
K	Na	Ca	Mg	Al	Zn	Fe	Sn	Pb	Cu	Ag	Au	Pt

fastest corrosion   slowest corrosion   silver, gold and platinum do not corrode at all

least resistant to corrosion   more resistant to corrosion   they are most resistant to corrosion

Now try this

target D-B

1. Explain which metal is more resistant to corrosion: lead or zinc. **(2 marks)**

2. Write a word equation to show the oxidation of copper **(1 mark)**

target B-A*

3. Write balanced equations to show:
   (a) the reduction of tin dioxide (SnO_2) using carbon **(2 marks)**
   (b) the oxidation of tin. **(2 marks)**

Recycling metals

When a metal object has reached the end of its life, the metal can be melted down and made into something new so it can be used again. The metal is recycled. Recycling metals instead of throwing them away has many benefits.

 Recycling metals means the Earth's supply of metals will last longer.

 If metals are recycled, we need fewer mines. This helps the environment, because mining causes dust and noise pollution, and damages the landscape.

 Recycling produces less pollution. Extracting some ores (such as lead) can produce sulfur dioxide. This does not happen when the metal is recycled.

 Less land is needed for landfill sites.

 Carbon dioxide is emitted when fossil fuels are used to heat ores or to generate the electricity used in electrolysis. Far less carbon dioxide is emitted by the recycling process, as less energy is needed. Carbon dioxide is a greenhouse gas (you will learn more about climate change on page 58).

 For most metals, it takes less energy to melt down used metals than it does to extract the metal from its ore. This makes some recycled metals more sustainable and cheaper.

Not all metals are suitable for recycling. For some metals it costs more to collect, sort and transport the used items than is saved by recycling them.

> **EXAM ALERT!**
>
> You need to give specific examples in an exam. Saying that recycling is 'better for the environment' is not detailed enough.
>
> Students have struggled with this topic in recent exams – **be prepared!**   ResultsPlus

Uses of metals

All metals conduct heat and electricity. They are malleable (can be hammered into shape) and ductile (can be stretched into wires). Not all metals have exactly the same properties. The uses of a particular metal depend on its properties.

Used for: jewellery because it looks attractive and does not corrode some electrical connections inside electronic devices because it is one of the best electrical conductors (it is too expensive to use for cables)	Used for: bridges, cars and buildings, because it is strong and resistant to corrosion.
gold | **steel** **copper** | **aluminium**	
Used for: electrical cables because it is a very good conductor of electricity water pipes because it does not corrode easily	Used for: drinks cans because it does not corrode aeroplanes and some cars because it has a low density

This is because copper is not very reactive.

You would not get a mark in an exam if you wrote that aluminium is 'light'. You need to say that it has a low density, or that it is 'lightweight'.

> **Now try this**

 target E-C

1. Explain why not all metals are recycled. **(2 marks)**

 target B-A

2. Describe two ways in which recycling lead can reduce pollution in the atmosphere. **(4 marks)**

Alloys

An alloy is a mixture of metals, and sometimes carbon. Steel is an alloy of iron with small amounts of other metals and some carbon. Steel is stronger than iron and more resistant to corrosion.

EXAM ALERT!

The particles shown in red in the diagram are particles of a different metal or of carbon. They are not 'alloy particles'. It is important to use the correct scientific words in your exam.

Students have struggled with questions similar to this in recent exams - **be prepared!**

ResultsPlus

Worked example

Use the diagrams to help you to explain why alloys are generally stronger than pure metals.

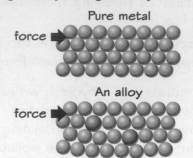

Pure metal

force ➤

An alloy

force ➤

In a pure metal the layers of atoms can slide over each. An alloy is stronger, because the atoms of the added metal (or carbon) are different sizes and they jam up the structure and stop the layers from sliding.

Gold

Pure gold is too soft to be useful. It is alloyed with other metals to make it stronger and harder. The percentage of gold in a gold alloy is measured in carats or described by a fineness. The fineness is the parts per thousand of gold in the alloy.

Percentage of gold alloy

	100%	75%	50%	37.5%
carats	24	18	12	9
fineness	1000	750	500	375

increasing strength ➤

Shape memory alloys

Smart materials have different properties in different conditions. Shape memory alloys are smart materials, as their shape changes when the temperature changes. Nitinol is an alloy of nickel and titanium. If the shape of something made from nitinol is changed, it will return to its original shape when it is heated.

Shape memory alloys are also used to make spectacle frames. These alloys are examples of new materials developed to meet new needs.

Worked example

Describe one use of nitinol.

It is used to keep arteries open. A stent is made from a tube of nitinol, which is cooled and squashed, so it is small enough to slide into the blocked artery. When the nitinol warms up to body temperature it resumes its original shape and holds the artery open.

Now try this

1. Use ideas about atoms in metals to explain why steel is stronger than iron. **(4 marks)**

2. A gold alloy is 62.5% gold. State the fineness and suggest its carat rating. **(2 marks)**

3. Explain why shape memory alloys are classed as smart materials. **(3 marks)**

Chemistry extended writing 2

Worked example

The main ore for aluminium is bauxite. Most deposits of bauxite occur near the surface of the Earth and are extracted by open-cast mining.

Discuss the economic and environmental advantages of recycling aluminium, including its effects on mining and supplies of ore. **(6 marks)**

Sample answer 1

Aluminium is used to make drinks cans. It is easy to collect to recycle, and can easily be sorted from iron cans using magnets. Many councils collect aluminium cans, and it saves them money because otherwise the aluminium would just get thrown away. Recycling is environmentally friendly, because it stops the cans being thrown away and littering the countryside.

This is a basic answer. All this is correct, but almost none of it answers the question that was asked. This question started with some information about how aluminium ore is mined, so writing about that should form part of your answer.

Sample answer 2

Open cast mines spoil the landscape and destroy habitats. They also produce noise and dust pollution. If aluminium is recycled, less ore needs to be extracted from the ground and so there will be less damage to the landscape. There will be less carbon dioxide released by vehicles used to transport and extract the ore as fewer of them will be needed. Also, supplies of the ore will last longer if they are not being used as quickly.

Aluminium is extracted from its ore using electrolysis. This uses a lot of energy. Most of the electricity is generated in power stations that burn fossil fuels. This puts carbon dioxide into the atmosphere, which is leading to climate change. Sulfur dioxide is also put into the atmosphere by power stations in some countries, and this can cause acid rain.

Recycling aluminium uses much less electricity than extracting it form ore. This makes recycled aluminium cheaper than 'new' aluminium. It also means that less fossil fuels need to be burnt to make the electricity, and so less carbon dioxide will be put into the atmosphere.

This is an excellent answer. It has explained both economic and environmental advantages of recycling aluminium.

Now try this

1. Some metals are found uncombined with other elements, but others have to be extracted from ores. Some methods of extraction are more expensive than others. Explain why different methods need to be used for different metals. Your answer should include examples. **(6 marks)**

You might need to look at the reactivity series on page 47.

Chemistry extended writing 3

Worked example

The properties of an alloy are different to the properties of the metal from which it is made. Explain, in terms of their structure, how alloy steels are more useful than pure iron. Include a diagram as part of your answer.

(6 marks)

Sample answer 1

Steel is made from iron. It is stronger than iron because the metal has some particles of other substances in it. When a force is applied to a piece of iron the iron can change shape because the layers of particles in it can slide over each other. In steel, the different sized particles stop the layers sliding. This makes the steel harder and stronger than the iron.

This is a basic answer. What is written is correct, but it does not mention the increased resistance to corrosion of some steel alloys, and it does not include a diagram. You can never get full marks unless you answer all of the question.

Sample answer 2

An alloy is a metal that is mixed with small amounts of another metal. Iron is made into steel to make it stronger. When a force is applied to pure iron, layers of atoms in the metal can slide over one another easily because they are all the same size. The metals added to make iron into steel have atoms of a different size. These atoms make it more difficult for the layers to slide. This means it takes a bigger force to make the layers slide, and the alloy is harder and stronger than the metal.

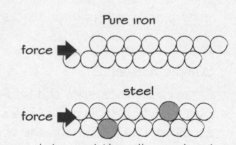

Steel is used for buildings, bridges and machines because it is stronger than iron. Iron goes rusty very easily when it is exposed to air and water. Some steels are resistant to corrosion, and stainless steels do not rust at all. This makes these alloys very useful for things such as cutlery.

This is an excellent answer. It explains why alloys are usually stronger and harder than the pure metals, and explains why steels are more useful than pure iron. It also includes a diagram that was asked for in the question.

Now try this

1. Containers of hydrochloric acid often have this symbol on them.

   Explain why this symbol is put onto containers of acid, and how you could neutralise spilled hydrochloric acid. Include balanced symbol equations for any reactions you mention.

   (6 marks)

Crude oil

Crude oil is a fossil fuel. It is a complex mixture of many different kinds of hydrocarbon molecules.

Hydrocarbons

Hydrocarbons are a group of compounds containing only hydrogen and carbon atoms. Each hydrocarbon molecule consists of a chain of carbon atoms surrounded by hydrogen atoms. All the atoms in each molecule are chemically joined together.

Hydrocarbons are a type of *compound*. The hydrogen and carbon *atoms* are chemically joined to make *molecules*. Students often lose marks by calling hydrocarbons a 'mixture' of carbon and hydrogen, or by saying that hydrocarbons are made of hydrogen and carbon molecules.

Worked example

Put an X in the boxes next to the two hydrocarbons.

☐ C_2H_6O
☒ C_5H_{10}
☒ C_2H_6
☐ C_2H_5OH

There are many different types of hydrocarbon, with different numbers of carbon atoms in them. The longest molecules can have hundreds of carbon atoms in them.

EXAM ALERT!

If you are asked to explain what is meant by 'hydrocarbon', remember to say that hydrocarbon molecules contain *only* hydrogen and carbon atoms.

Students have struggled with exam questions similar to this in recent exams - **be prepared!**

Results Plus

Fractional distillation

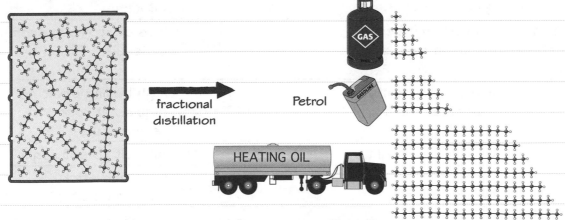

fractional distillation

Petrol

GAS

GASOLINE

HEATING OIL

Crude oil contains a mixture of different hydrocarbon molecules. This mixture is not very useful.

These are more useful fractions. Each fraction is a simpler mixture than crude oil.

Now try this

target D-C

1. (a) Give the name of the process used to separate crude oil into fractions. **(1 mark)**
   (b) Explain why crude oil is split into fractions. **(2 marks)**

target C-B

2. Write the formula of the long-chain hydrocarbon shown in the top diagram. **(1 mark)**

3. Explain why the fractions obtained in fractional distillation are still referred to as 'mixtures'. **(2 marks)**

Crude oil fractions

Each fraction obtained from crude oil contains a mixture of a few different hydrocarbons. Each fraction has different properties and different uses.

Worked example

Complete the table to show the uses of the different fractions.

Fraction	Uses
gases	heating and cooking in homes
petrol	fuel for cars
kerosene	fuel for aircraft
diesel oil	fuel for some cars and trains
fuel oil	fuel for large ships and in some power stations
bitumen	surfacing roads and roofs

The properties of the different fractions change gradually, in the order shown in the table. You need to remember the order of the fractions, as well as their names.

You need to remember the names of the different fractions, and their uses.

Properties

This table shows how different properties of fractions change from the gases to bitumen.

Property	gas	bitumen
number of carbon and hydrogen atoms in the molecules	smallest number of atoms in molecules	lots of atoms in each molecule
boiling point	low boiling point (gases at room temperature)	high boiling point (liquids at room temperature)
ignition	easy to set alight	difficult to set alight
viscosity	runny (low viscosity)	thick and sticky (high viscosity)

Now try this

target C-B

1. Describe the similarities and differences between the fractions used as fuel for cars and the fraction used as fuel for ships.   **(4 marks)**

target C-A

2. Suggest why bitumen is not used as a fuel.   **(2 marks)**

When a question asks you to 'suggest' something, it usually means you have to apply your knowledge – you need to think about the facts you have and how they apply to the question.

Combustion

The scientific name for burning is combustion. When hydrocarbons burn they combine with oxygen to produce carbon dioxide and water. The reaction releases energy.

Combustion of methane

Methane is a hydrocarbon. It is the main gas in natural gas.

methane + oxygen → carbon dioxide + water
$$CH_4(g) + 2O_2(g) \rightarrow CO_2(g) + 2H_2O(l)$$

The methane has been oxidised. Combustion is an example of an oxidation reaction.

Students often lose marks by forgetting that water is produced when hydrocarbons burn. You can't normally see the water because the heat of the reaction means it is produced as a gas. The state symbol in the equation is (l), as water is a liquid at room temperature.

Complete combustion

A combustion reaction is described as complete combustion if all the hydrocarbon is oxidised and the only products are carbon dioxide and water. Complete combustion happens when there is plenty of oxygen available.

If there is not enough oxygen available, then incomplete combustion occurs (see next page).

Worked example

Label the diagram and use it to help you to explain the test for carbon dioxide.

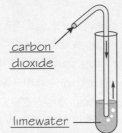

carbon dioxide

limewater

When carbon dioxide is bubbled through limewater, the carbon dioxide combines with the limewater to form solid calcium carbonate. This makes the limewater turn milky.

EXAM ALERT!

This is a good answer because it explains the test *and* the result you will see if the gas is carbon dioxide.

The test for carbon dioxide is not that it will put out a lighted splint. Carbon dioxide *will* put out a lighted splint, but lots of other gases do this too so it is NOT a good test for carbon dioxide.

Students have struggled with exam questions similar to this in recent exams - **be prepared!**   ResultsPlus

Now try this

target E-C

1. A reaction produces a gas that you think is carbon dioxide.
   Describe how you would show that the gas is carbon dioxide.   **(2 marks)**

target D-A*

2. Propane (C_3H_8) is a hydrocarbon that is part of the gases fraction from crude oil.

   (a) Write a word equation for the complete combustion of propane.   **(1 mark)**
   (b) State which substance is oxidised in this reaction.   **(1 mark)**
   (c) Write a balanced equation for this reaction. Include state symbols.   **(3 marks)**

Incomplete combustion

In complete combustion, all the carbon atoms in a hydrocarbon are oxidised to form carbon dioxide. All the hydrogen atoms that were in the hydrocarbon are oxidised to form water.

Incomplete combustion

Sometimes there is not enough oxygen available to allow all these oxidation reactions to take place. When incomplete combustion takes place all the hydrogen atoms become oxidised to form water, but the carbon atoms may form:

- some carbon dioxide (CO_2)
- some carbon monoxide (CO)
- some soot (particles of solid carbon).

    methane + oxygen → carbon dioxide + carbon monoxide + carbon + water

Different amounts of these substances are produced, depending on how much oxygen is available. Incomplete combustion reactions release energy.

Cars and boilers

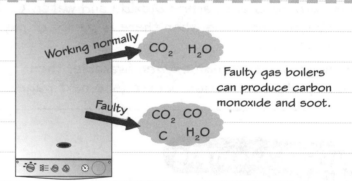

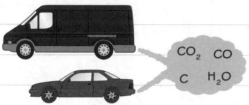

There is always incomplete combustion in vehicle engines.

Faulty gas boilers can produce carbon monoxide and soot.

Worked example

Explain the problems caused by:
(a) carbon monoxide   (b) soot.

(a) Carbon monoxide is a toxic gas. It reduces the amount of oxygen that the blood can carry around the body. Breathing in carbon monoxide can kill you.

(b) Soot can:
- build up in chimneys and eventually cause fires
- cause lung diseases
- make buildings dirty.

Dirty marks around gas appliances show that soot is being formed. This shows that incomplete combustion is taking place. Carbon monoxide may also be forming, but the soot does not directly show that this is happening.

⬅ You can use bullet points to help you to organise your answers.

Now try this

target
D-B

1. Explain why carbon monoxide harms the body. **(2 marks)**

target
C-A

2. Describe the similarities and differences between complete and incomplete combustion. **(3 marks)**

3. Sam says that soot marks around a boiler show that carbon monoxide is being produced. Explain what is wrong with this statement. **(2 marks)**

Acid rain

Acid rain has a lower pH than normal rain.

Sulfur dioxide

Most hydrocarbon fuels contain impurities such as sulfur. When the hydrocarbons burn, the sulfur is oxidised to form sulfur dioxide.

Liquid hydrocarbon fuels such as petrol and diesel contain sulfur. Coal also contains sulfur. Natural gas and the gases produced from crude oil do not contain sulfur.

Sulfur dioxide dissolves in water and makes the water **acidic**.

Worked example

Explain what acid rain is and how is it formed.

Rain water is naturally slightly acidic because it contains dissolved carbon dioxide and other acidic gases normally in the air. Acid rain is rain that is more acidic than normal because sulfur dioxide has dissolved in it.

The effects of acid rain

sulfur dioxide dissolves in water in the air

waste gases from power stations and vehicles contain sulfur dioxide

acid rain speeds up the weathering of buildings and statues

trees are damaged

rain is more acidic than normal

rivers, lakes and soils are more acidic, which harms organisms living in them

Cleaning up

The problem of acid rain is being reduced in Europe and North America by:

* removing sulfur from petrol, diesel and fuel oil
* removing acidic gases from power station emissions.

EXAM ALERT!

Make sure you read all the information in a question carefully. A recent exam question asked how passing waste gases from power stations through calcium carbonate helps to reduce the amount of acid rain. Many students said that the calcium carbonate was added to the clouds, even though the question told them how the calcium carbonate was used!

Students have struggled with this topic in recent exams – **be prepared!**   Result Plus

Now try this

1. Explain why acid rain speeds up the erosion of limestone rocks and buildings. Include a word equation in your answer. **(5 marks)**

2. Alex says that acid rain is rain with a pH less than 7. Comment on this statement. **(4 marks)**

Climate change

Keeping warm

The Earth is warmed by the Sun, and loses heat to space. Some of the gases in the atmosphere trap heat and help to keep the Earth warm. These gases include carbon dioxide, methane and water vapour. Without these gases the average temperature on Earth would be about −18°C. These gases are sometimes called 'greenhouse gases'.

Carbon dioxide

The temperature of the Earth varies naturally over long periods of time. The amount of carbon dioxide in the atmosphere also varies naturally, and this is linked with the temperature changes.

Since 1800 the proportion of carbon dioxide in the atmosphere has increased because humans have been burning fossil fuels and so releasing carbon dioxide into the atmosphere. Farming also adds methane to the atmosphere. The graph below shows that the Earth's temperature and the concentration of carbon dioxide have both risen. This warming is called climate change.

Worked example

Evaluate how far the graph provides evidence for climate change.

The graph shows that both temperature and carbon dioxide concentration have been rising since 1950. There is a correlation but this does not <u>prove</u> that rising carbon dioxide levels are causing the warming.

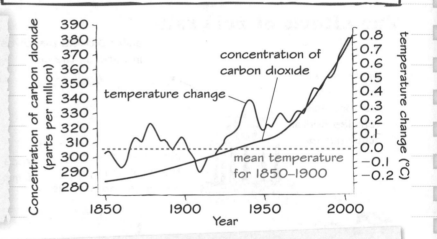

Most scientists agree that carbon dioxide is the main cause of global warming, but this opinion is based on computer modelling and other information, not just on graphs like this.

Reducing CO₂

Chemists are trying to find ways to control the amount of CO_2 in the atmosphere. Two methods being investigated are:

- 'seeding' the oceans with iron compounds. This might encourage microscopic plants to grow, which would use carbon in photosynthesis. This carbon would eventually be incorporated into shells and end up in sediments.
- capturing carbon dioxide from power station waste gas and converting it into hydrocarbons to be used as fuels.

Now try this

target
D-B

1. (a) Name three gases that help to keep the Earth warm. **(3 marks)**
   (b) Explain why the levels of two of these gases are increasing. **(2 marks)**
2. Explain two ways in which it may be possible to control the amount of carbon dioxide in the atmosphere. **(4 marks)**

Biofuels

Biofuels are fuels obtained from living organisms and can be used instead of fossil fuels. They are a renewable resource, because more plants can be grown to replace the ones used as fuel.

Burning biofuels

Biofuels include...

- plants grown to be burned, such as wood
- sugar cane or sugar beet, which can be converted to ethanol. This can be used instead of petrol, and reduces the demand for petrol.

Burning biofuels adds less carbon dioxide to the atmosphere than burning fossil fuels. This is because the plants that were used to make the biofuel absorbed carbon dioxide from the atmosphere when they grew.

Carbon neutrality

A carbon neutral fuel is one that does not increase the total amount of carbon dioxide in the atmosphere when it burns.

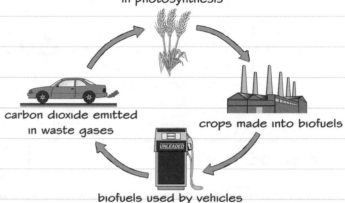

carbon dioxide absorbed in photosynthesis

carbon dioxide emitted in waste gases

crops made into biofuels

biofuels used by vehicles

Worked example

Explain why most biofuels are not carbon neutral.

When making biofuels, energy is needed to fertilise, harvest and transport the crops, and to make them into a fuel that can be used in vehicles.

At the moment this energy usually comes from burning fossil fuels, so some carbon dioxide is added to the atmosphere when the biofuels are produced.

 This is a good answer because it explains why energy is needed to produce biofuels.

Advantages and disadvantages

✓ Biofuels are renewable.

✓ Biofuels add less carbon dioxide to the atmosphere than burning fossil fuels.

✗ Growing crops to make into biofuels means that less land is available for growing food. This leads to higher food prices and may lead to some people being short of food.

Now try this

 target **D–C**

1. Wood and ethanol are both biofuels. Describe their similarities and differences.   **(4 marks)**

2. Explain two advantages of biofuels compared with fossil fuels.   **(4 marks)**

3. Explain why biofuels are not normally carbon neutral.   **(3 marks)**

Choosing fuels

Petrol, kerosene and diesel oil are non-renewable fossil fuels. They are all obtained from crude oil. Methane is a non-renewable fossil fuel found in natural gas.

Hydrogen is also used as a fuel. In rockets, hydrogen is burnt to release energy. Hydrogen can also be used in cars and other vehicles by using a fuel cell.

Worked example

Explain what a hydrogen fuel cell does.

A fuel cell combines hydrogen and oxygen to form water without burning. This reaction releases energy in the form of electricity.

Remember the similarities and differences between burning hydrogen and using it in a fuel cell:
- both reactions react hydrogen with oxygen to form water
- burning hydrogen releases energy as heat
- using hydrogen in a fuel cell releases energy as electricity – the hydrogen does not burn in the fuel cell.

A good fuel...

- should burn easily
- should not produce ash or smoke
- should release a lot of heat energy
- should be easy to store and transport.

Investigating fuels

If you are given a range of fuels you can work out which one gives off most heat by measuring the time taken for each fuel to raise the same volume of water by 50 °C. Don't forget to measure the mass of the fuel before and after the experiment to find out how much you have used. There are lots of other variables to control, such as volume of water.

Hydrogen versus petrol

Petrol	Hydrogen
✓ burns easily	✓ burns easily
✓ does not produce ash or smoke	✓ does not produce ash or smoke
✗ produces carbon dioxide and carbon monoxide as well as water when it burns	✓ only produces water when it burns
✓ releases more energy per kg when it burns than fuels such as coal or wood	✓ releases nearly three times as much energy per kg as petrol
✓ is a liquid, so it is easy to store and transport	✗ is a gas, so it has to be stored at high pressure
	✗ filling stations would need to be adapted for hydrogen to be used in cars

Now try this

target D-B

1. Describe the similarities and differences between petrol and methane. **(4 marks)**

target D-B

2. Explain why petrol is a better fuel for cars than coal. **(2 marks)**

Alkanes and alkenes

Alkanes and alkenes are two groups of hydrocarbon molecules.

Alkanes

Crude oil is made of a mixture of different alkane molecules. The three simplest alkanes are methane, ethane and propane. These compounds are found in natural gas.

Alkanes are saturated hydrocarbons. This means that all the bonds joining carbon atoms to each other are single bonds.

Alkenes

Alkenes are similar to alkanes, except that alkene molecules have at least one double bond between two of the carbon atoms. They are unsaturated molecules.

It is very easy to get alkanes and alkenes confused, as the names are very similar. Read a question very carefully to check whether the name has an a or an e in the middle!

Names

The names of all alkanes end in '-ane' and the names of all alkenes end in '-ene'. The first part of the name tells you the number of carbon atoms in each molecule.

- methane (CH_4) has only one carbon atom
- ethane (C_2H_6) and ethene (C_2H_4) have two carbon atoms
- propane (C_3H_8) and propene (C_3H_6) have three carbon atoms

Structures

You need to be able to draw the structures of all the molecules mentioned on this page.

methane   ethane   propene

The bromine test

The bromine test is used to find out if a liquid contains double bonds. Bromine water (bromine dissolved in water) is an orange liquid but becomes colourless when mixed with unsaturated molecules.

Bromine water can also be pale yellow or red-brown, depending on how much bromine is dissolved. The important thing to remember is that bromine water is **coloured**.

EXAM ALERT!

When writing about the colour of bromine water remember that bromine water is clear and orange. It changes to clear and colourless when a substance with double bonds is added. Remember that clear means transparent and is not a colour!

Students have struggled with this topic in recent exams – **be prepared!**   Results**Plus**

Now try this

target D-B

1. Draw the structure of the following:
   - (a) propane   **(3 marks)**
   - (b) ethene.   **(3 marks)**

2. (a) Describe two similarities between ethene and ethane.   **(2 marks)**
   - (b) Describe one difference.   **(1 mark)**

3. Explain what you would see if you shake up bromine water with the following:
   - (a) ethane   **(2 marks)**
   - (b) propene.   **(2 marks)**

Cracking

Crude oil contains a mixture of long and short alkane molecules, but the short ones are the most useful. The long chains are broken down into shorter chains by cracking.

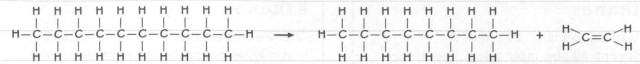

The long molecules are not very useful.

Cracking breaks down the long molecules by heating them.

Cracking produces shorter-chain alkanes, which are useful as fuels.

Cracking also produces alkenes, which are used to make polymers.

Remember that alkanes are saturated molecules (they have only single bonds), and alkenes are unsaturated molecules (they contain a double bond).

EXAM ALERT!

Remember the difference between fractional distillation and cracking. Fractional distillation is a physical separation process, and does *not* change any molecules. Cracking is a chemical reaction, and *breaks down* the molecules.

Students have struggled with this topic in recent exams – **be prepared!**   Results Plus

Cracking paraffin

Paraffin is an alkane. Liquid paraffin can be cracked in the laboratory using the apparatus shown in the diagram below.

- The porous pot is heated strongly.
- The liquid paraffin is heated and evaporates.
- The paraffin vapour passes over the hot porous pot and the hydrocarbon molecules break down.
- One of the products is ethene, which is a gas and collects in the other tube.

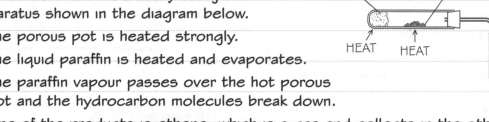

liquid paraffin on mineral wool
broken porous pot
delivery tube
ethene gas
HEAT   HEAT
water

Supply and demand

The graph shows the difference between supply and demand for different fractions in crude oil. This shows why cracking is necessary.

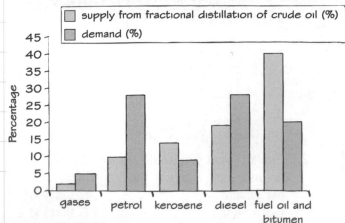

Now try this

target C-B

1. Explain why some alkanes are cracked, referring to the bar chart in your answer. **(4 marks)**

target C-A*

2. (a) Suggest why the upside-down test tube full of water is used when cracking paraffin in the laboratory. **(2 marks)**

   (b) Suggest why the first few bubbles of gas should be allowed to escape and not be collected. **(2 marks)**

Polymerisation

Ethene molecules have a double bond between the two carbon atoms. Lots of ethene molecules can be made to join up to form one long molecule called poly(ethene). The (poly)ethene only has single bonds between carbon atoms.

Many (n) ethane monomers polymerise (react together) to form a long chain of poly(ethene) with n repeat units.

Other small alkene molecules can be joined up to make long molecules. The small molecules are called monomers. Lots of monomers join up to make polymers.

Worked example

Fill in the gaps to show the names of the polymers made from these monomers.

Propene monomers make <u>poly(propene)</u>.

Chloroethene monomers make <u>poly(chloroethene)</u>.

Tetrafluoroethene monomers make poly(tetrafluoroethene) or <u>PTFE</u>.

Poly(chloroethene) is also called PVC.

Poly(tetrafluoroethene) is also called PTFE. You don't need to memorise 'poly(tetrafluoroethene)' – just remember the initials.

Properties and uses

Different polymers have different properties, so they have different uses.

Polymer	Properties	Uses
poly(ethene)	flexible, cheap, good electrical insulator	plastic bags, plastic bottles and clingfilm
poly(propene)	flexible, shatterproof, has a high softening point	buckets and bowls
poly(chloroethene) (PVC)	tough, cheap, long lasting, good electrical insulator	window frames, gutters and pipes, insulation for electrical wires
PTFE	tough, slippery, resistant to corrosion, good electrical insulator	non-stick coatings for frying pans and skis, containers for corrosive substances, stain-proofing carpets, insulation for electric wires

Now try this

target E-C

1. Explain which properties of PTFE make it suitable for use on frying pans.
(2 marks)

target D-B

2. Describe the difference between a monomer and a polymer. **(4 marks)**

target B-A*

3. The diagram shows a molecule of chloroethene.

(a) Write down the formula for chloroethene. **(1 mark)**

(b) Write an equation for the formation of poly(chloroethene) (PVC) similar to the one at the top of the page. **(3 marks)**

Problems with polymers

Biodegradability

Many materials we throw away are biodegradable. They eventually rot away because microbes feed on them and break them down. Most polymers are not biodegradable. This is useful, because products made from them last a long time. It is also a problem, because when polymers have been thrown away they do not rot.

Don't get 'biodegradable' mixed up with 'corrosion'. Materials biodegrade when microbes decompose them. Materials corrode when they are attacked by chemical substances in the environment.

Disposing of polymers

Landfill sites
- polymers are not biodegradable
- they last for many years
- we are running out of landfill sites ✗

Burning ✗
- many polymers release toxic gases when they burn

Disposing of polymers

Recycling
- melting or breaking down polymers to make new objects ✓

Biodegradable polymers
- these are being developed
- they will rot away in landfill sites ✓

Worked example

Explain why recycling polymers is harder than recycling glass.

There are many different kinds of polymer. The polymer waste has to be sorted before the different polymers can be broken down or melted and made into new objects.

Watch out! Reusing and recycling are not the same. Reusing means using the same object several times (such as using plastic carrier bags for shopping many times). Recycling means that the material is made into new objects.

Now try this

target **C-A**

1. Explain the advantages and disadvantages of burning and landfill as ways of disposing of waste polymers.
(4 marks)

target **C-A***

2. Suggest some advantages and disadvantages of introducing biodegradable polymers. **(4 marks)**

You need to start by describing the property that makes biodegradable polymers different from other polymers.

Chemistry extended writing 4

Worked example

One possible way of reducing the amount of carbon dioxide in the atmosphere is to convert some of it into hydrocarbons, which can then be used as fuels.

Explain why this is necessary, and discuss how effective it will be in reducing the concentration of carbon dioxide in the atmosphere. **(6 marks)**

Sample answer 1

Carbon dioxide is a greenhouse gas and it helps to keep the Earth warm. Levels of carbon dioxide have been increasing as humans burn lots of fossil fuels and this has been linked to climate change. If we can take some of the carbon dioxide out of the air then the Earth might not warm up quite so much. Converting some carbon dioxide to hydrocarbons will help to do this.

This is a basic answer. There is a good explanation for why it is a good idea to remove some carbon dioxide from the atmosphere, but there is no discussion about how effective it might be to try to do this by using carbon dioxide to produce hydrocarbons.

Sample answer 2

Carbon dioxide traps heat in the atmosphere and it is needed to keep us warm enough, but there is getting to be too much of it because of humans burning fossil fuels and too much of it will make the Earth warmer. This is called climate change, and is not a good thing so it is good to take some of it out again to make into fuels. CO_2 from power stations gets trapped to make the fuel but when it burns the CO_2 will go back into the air so you could say that it won't make a difference but if it was not burnt then new petrol from the ground would be burnt instead so overall it is better to use it to make fuels than not to.

This is an excellent answer from the point of view of the science. However it would not gain full marks because the presentation is not very good.

Now try this

1. Describe cracking and explain its economic benefits. Refer to information from the bar chart as part of your answer. **(6 marks)**

Composition of crude oil and the demand for the different fractions.

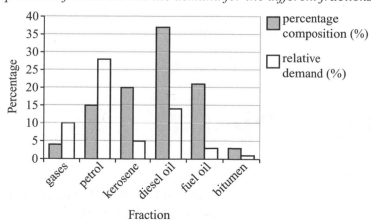

Chemistry extended writing 5

Worked example

Hydrocarbon fuels release energy when they burn. Two students tested the hypothesis that the amount of energy released per gram of fuel depends on the number of carbon atoms in the molecules. The graphs show their results.

Suggest how the two students may have carried out their investigation, and discuss which student has probably carried out the best test of the hypothesis. **(6 marks)**

Sample answer 1

They could burn each fuel for the same time, and put a beaker of water over the flame. They weigh the fuel and the burner before and after each test so they can work out how much fuel is used. They get the data for the graph by measuring the temperature rise in the water and dividing this by the mass of fuel burnt.

They would have to use the same volume of water each time, at the same starting temperature. They should use the same size and shape of beaker. The fuel burner would have to be the same distance from the beaker as well.

This is a good answer. There is a good description of how the variables could be controlled so that the results can be compared. However, this student has not discussed the results at all, nor compared the two graphs.

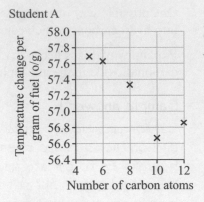
Student A

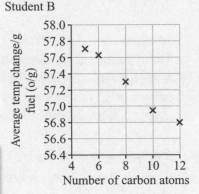
Student B

Sample answer 2

This answer gave a description of the method and fair testing similar to Sample answer 1, but then added the following:

Both graphs show the temperature change decreasing as the number of carbon atoms in the molecules increases, so they show the hypothesis is correct (as long as a fair test was carried out). Student A's result for the fuel with 10 carbon atoms looks as if it might be wrong. It might be that this student only tested each fuel once. Student B has plotted an average so they probably tested each fuel more than once. Doing it more than once would have let them see if they had made any mistakes, and so I think Student B did a better test of the hypothesis.

This is an excellent answer. As well as describing the method and showing how the variables can be controlled (not shown above), the student has said whether or not the results support the hypothesis, and also discussed which student did the best test.

Now try this

1. Discuss the opinion shown.

   If we are running out of oil, we shouldn't be wasting it by using it to make plastic bottles! We should go back to glass milk bottles. **(6 marks)**

The Solar System

Many early people thought that the Sun and all the planets moved around the Earth.

Changing ideas

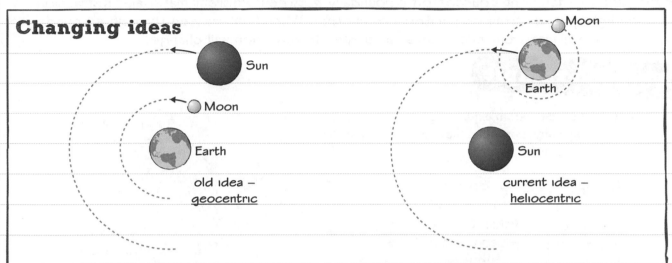

old idea – <u>geocentric</u>

current idea – <u>heliocentric</u>

In about 130 CE a Greek astronomer called Ptolemy published a geocentric ('Earth-centred') model that tried to explain the motions of the planets.

In 1543 Nicolaus Copernicus published a book that argued that the Earth and the planets were in orbits around the Sun. This is a heliocentric ('Sun-centred') model.

Ways of observing the Universe using light waves

- We can see some stars and planets using the naked eye.
- Telescopes allow us to see more distant and fainter objects, because the large lenses or mirrors gather more light.
- Photographs provide a record of observations that can be studied later or analysed by computer. Cameras can also record light arriving through a telescope over many minutes, so they can produce pictures of objects too dim to see just by looking through a telescope.

Galileo

Galileo Galilei was the first astronomer to make observations using telescopes. In 1610 he reported that he had observed four moons orbiting Jupiter. These were the first objects seen to be orbiting around something that was not the Earth. This observation supported Copernicus's ideas.

You don't need to remember the dates for Ptolemy, Copernicus and Galileo, but you should remember the order of the different ideas and observations.

Now try this

target E-C

1. (a) What is the difference between a geocentric and a heliocentric model of the Solar System?
    (2 marks)
    (b) What evidence did Galileo find that backed up Copernicus's ideas? **(1 mark)**
    (c) Discuss how this evidence supported the heliocentric model.
    (2 marks)

target D-B

2. Ancient astronomers knew about six planets in the solar system (including the Earth). The planet Uranus was not discovered until 1781 and Neptune was not discovered until 1846. Explain these statements. **(4 marks)**

Reflection and refraction

Light travels as waves, normally in straight lines unless it is reflected or refracted. Reflection is when light bounces off a boundary between different materials. Refraction is when light passes from one material to another. When light passes through a boundary between two transparent materials at an angle, its direction will change.

Worked example

Draw a labelled diagram explaining why waves change direction when they pass from air into glass or water.

The key points to show on the diagram are the change in direction towards the normal, and the decrease in wavelength as the wave moves into the medium where it travels more slowly.

The waves are travelling in this direction.

air

normal

This part of the wave slows down first.

glass or water

The waves end up travelling in this direction.

The waves travel more slowly in this medium.

Lenses

Lenses are specially shaped pieces of transparent material that change the direction of light rays. A converging lens makes rays of light come together.

- The point at which parallel rays are brought together is called the focal point. The distance between a lens and the focal point is called the focal length.
- The fatter the lens, the shorter the focal length.
- The magnification of the image depends on the focal length and the distance between the object and the lens.

A converging lens can be used as a magnifying glass if it is held close to an object.

This is a virtual image, because the rays of light only appear to be coming from it. The image is also magnified and the right way up.

Finding the focal length

Find the focal length of a converging lens by using it to focus parallel rays of light from a distant object onto a piece of paper. The rays of light will meet at the paper and form a real, inverted image. The focal length is the distance between the lens and the paper.

Now try this

1. (a) What is the focal length of a lens?
     (3 marks)
   (b) Describe how to measure the focal length of a converging lens. **(4 marks)**

2. Explain why waves refract away from the normal when they move into a medium in which they travel faster. **(4 marks)**

Telescopes

Simple telescope

The first telescopes used lenses to gather more light and to magnify the image. A refracting telescope uses two converging lenses. The objective lens brings rays of light from a distant object to a point to form an image. The eyepiece lens acts as a magnifying glass and magnifies this image.

parallel rays of light from a
distant object form a real image

user sees a virtual,
magnified image

eyepiece lens

objective lens

Reflecting telescope

A reflecting telescope uses a curved mirror to gather light from distant objects and a converging lens as an eyepiece to magnify the image. If a telescope needs to see very dim objects, it is easier to make a very large mirror than a very large lens. Reflecting telescopes give better quality images than refracting telescopes of similar sizes.

Worked example

Complete the labels on this diagram of a reflecting telescope.

You need to be able to explain how the two different types of telescope work, but you will not be asked to draw ray diagrams.

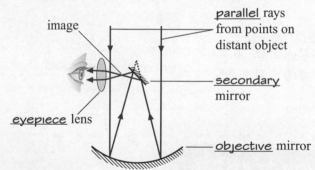

image

<u>parallel</u> rays from points on distant object

<u>secondary</u> mirror

<u>eyepiece</u> lens

<u>objective</u> mirror

The eyepiece lens <u>magnifies</u> the <u>image</u> produced by the <u>mirrors</u>.

Now try this

target
D-C

1. Which part of these telescopes is responsible for magnifying the image?
   (a) a simple telescope   **(1 mark)**
   (b) a reflecting telescope   **(1 mark)**

target
C-B

2. Each time light passes from one transparent material into a different transparent material, some of it is reflected. Explain why refracting telescopes reduce the amount of light passing through them more than reflecting telescopes.   **(4 marks)**

Waves

Waves transfer energy and information without transferring matter.

Waves can be described by their

- frequency – the number of waves passing a point each second, measured in hertz (Hz)
- speed – measured in metres per second (m/s)
- wavelength and amplitude

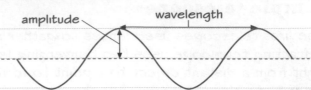

Watch out! Remember that the amplitude is *half* of the distance from the top to the bottom of the wave. Students often get this wrong.

Longitudinal waves

Sound waves and seismic P waves are longitudinal waves. The particles in the material the sound is travelling through move back and forth along the same direction that the sound is travelling.

longitudinal wave

air particles move like this

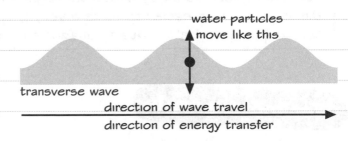

direction of wave travel
direction of energy transfer

Particles in a longitudinal wave move along the same direction as the wave.

Transverse waves

Waves on a water surface, electromagnetic waves and seismic S waves are all transverse waves. The particles of water move in a direction at right angles to the direction the wave is travelling.

water particles move like this

transverse wave

direction of wave travel
direction of energy transfer

Particles in a transverse wave move across the direction the wave is travelling.

Now try this

target D–C

1. (a) Sketch a transverse wave and mark the amplitude and wavelength on it. **(3 marks)**
   (b) Draw an arrow to show which way the wave moves. **(1 mark)**
   (c) Draw a small particle on the wave, with arrows to show which way it moves. **(1 mark)**

2. The graph shows a wave. Each vertical square represents 1 mm. Work out the amplitude of the wave. **(1 mark)**

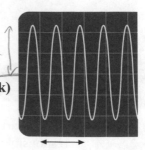

Wave equations

You may need to use one of these equations in your exam.

Speed, frequency and wavelength

$v = f \times \lambda$

v = wave speed (metres per second, m/s)

f = frequency (hertz, Hz)

λ = wavelength (metres, m)

λ is the Greek letter 'lambda'.

Speed, distance and time

$v = \dfrac{x}{t}$

v = wave speed (metres/second, m/s)

x = distance (metres, m)

t = time (seconds, s)

Worked example

A seismic wave has a frequency of 15 Hz and travels at 4050 m/s. Calculate its wavelength.

$$\lambda = \frac{v}{f}$$

$$= \frac{4050 \text{ m/s}}{15 \text{ Hz}}$$

$$= 270 \text{ m}$$

 Choose the equation that includes frequency and wavelength. Rearrange it so that it can be used to calculate the wavelength. See page 102 for help with the formula triangle.

Worked example

A wave on the sea is travelling at 4 m/s. Calculate how long it takes to travel along a 20 m long pier.

$$t = \frac{x}{v}$$

$$= \frac{20 \text{ m}}{4 \text{ m/s}}$$

$$= 5 \text{ s}$$

You do not need to remember the equations. They will be provided on a formula sheet in your exam. But you do need to be able to choose the correct equation to use, and to use the correct units. You also need to be able to rearrange any of the equations on the formula sheet.

Now try this

 target **C-B**

1. A sound wave with a frequency of 100 Hz has a speed of 330 m/s. Calculate its wavelength. **(3 marks)**

 target **B-A**

2. A wave in the sea travels at 25 m/s. Calculate the distance it travels in one minute. **(4 marks)**

Beyond the visible

The light we can see with our eyes is visible light. Light can be split up into the seven colours of the visible spectrum. Visible light is part of a 'family' of electromagnetic waves (see opposite page).

The colours of the visible spectrum are...

Red	Orange	Yellow	Green	Blue	Indigo	Violet

longest wavelength   shortest wavelength
lowest frequency   highest frequency

> Remember the order of the colours by remembering the name Roy G Biv.

Herschel and infrared

William Herschel discovered infrared radiation.

He used a prism to split light into the colours of the visible spectrum.

He used a thermometer to investigate the heating effect from each colour.

The heating increased from violet to red.

He investigated the region beyond the red.

He found an even bigger heating effect. He had discovered infrared radiation.

> 'Infra' means 'below'. Infrared radiation is 'below' visible red light because it has a longer wavelength and lower frequency.

Ritter and ultraviolet

Worked example

Explain how Johann Ritter discovered ultraviolet radiation.

Ritter used a chemical called silver chloride, which turns black when light shines on it. Silver chloride turns black faster in violet light than it does in red light. It turned black even faster when it was exposed to invisible radiation beyond the violet part of the spectrum. Ritter had discovered ultraviolet radiation.

> When you are asked to explain a contribution that someone makes to science you need to explain what they did and then what they discovered.

> 'Ultra' means 'above'. Ultraviolet radiation is 'above' visible violet light because it has a shorter wavelength and higher frequency.

Now try this

target
D-C

1. Explain the effect on a thermometer of putting it into the ultraviolet part of the spectrum. **(3 marks)**

target
C-A

2. The waves in the visible spectrum have wavelengths between 380 and 750 nm, and frequencies between 400 and 790 THz.
   (a) State a wavelength and a frequency of red light. **(2 marks)**
   (b) Estimate a wavelength and a frequency of green light. **(2 marks)**

> 1 nanometre (nm) = 10^{-9} m, 1 terahertz (THz) = 10^{12} Hz – but you don't need to know this information to answer the question.

The electromagnetic spectrum

Infrared radiation, visible light and ultraviolet radiation are all part of the electromagnetic spectrum.

All electromagnetic waves...

- are transverse waves (the electromagnetic vibrations are at right angles to the direction the wave is travelling – see page 70)
- travel at the same speed in a vacuum.

Watch out! The different parts of the electromagnetic spectrum have different properties, which you will read about on the following pages. But it is important to remember that some of their properties are *the same*. They are *all* transverse waves, and *all* travel at the same speed in a vacuum.

The electromagnetic spectrum

The electromagnetic spectrum is a group of waves that have different wavelengths and frequencies that are part of a continuous spectrum. Scientists put the waves into different groups according to their properties.

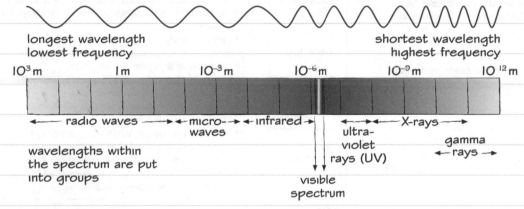

longest wavelength
lowest frequency

shortest wavelength
highest frequency

10^3 m   1 m 10^{-3} m 10^{-6} m 10^{-9} m 10^{-12} m

radio waves → ← micro-waves → ← infrared → ultra-violet rays (UV) → X-rays → gamma ← rays →

wavelengths within the spectrum are put into groups

visible spectrum

Worked example

List the parts of the electromagnetic spectrum, starting with the longest wavelength waves.

radio waves, microwaves, infrared, visible light, ultraviolet, X-rays, gamma rays

You can use a mnemonic to help you to remember the order. A mnemonic is a sentence or phrase whose words have the same initial letters as the list you are trying to remember. For example: Red Monkeys In Vans Use X-ray Glasses.

Now try this

1. An electromagnetic wave has a wavelength of 1 mm. State the type of electromagnetic wave.   **(1 mark)**

2. A radio wave with a frequency of 6×10^7 Hz has a wavelength of 5 m in a vacuum. Calculate the speed of this wave.   **(3 marks)**

3. Choose your answers to this question from the numbers below.

    6.2×10^8 m/s 5.25×10^8 m/s 3×10^8 m/s 2.25×10^8 m/s

State the speed of visible light in the following materials, and explain why you chose each answer:
(a) a vacuum   **(2 marks)**
(b) glass.   **(2 marks)**

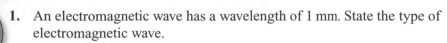

You need to use your answer to question 2 to help you.

Dangers and uses

All electromagnetic waves transfer energy. The amount of energy transferred depends on the frequency of the wave. The higher the frequency, the more energy the wave transfers. Different parts of the spectrum have different uses, which depend on their properties.

Uses

Short wavelength

Dangers

X-rays and gamma rays can cause mutations (changes) to the DNA in cells in the body. This may kill the cells or cause cancer.	Gamma rays	• to sterilise food and medical equipment • in scanners to detect cancer • to treat cancer

X-rays

Gamma rays

- to look inside objects, including human bodies
- in airport security scanners, to see what people have in their luggage

UV in sunlight can damage skin cells, causing sunburn. Over time, exposure to UV can cause skin cancer. UV can also damage the eyes.

ultraviolet

- to detect security marks made using special pens
- inside fluorescent lamps
- to detect forged banknotes (real banknotes have markings that glow in UV light)
- to disinfect water

visible light

- allows us to see, lights up rooms, buildings and roads
- photography

Infrared radiation transfers heat energy. Too much infrared radiation can cause skin burns.

infrared

- in cooking (by grills and toasters)
- to make thermal images (images using heat), used by police and rescue services
- in short range communications, such as between laptops or other small computers
- in remote controls for TVs and other appliances, where the signal only has to travel short distances
- to send information along optical fibres
- in security systems such as burglar alarms, to detect people moving around

Microwaves heat water – so they can heat the water inside our bodies. Heating cells can damage or kill them.

microwaves

radio waves

- in mobile phones, and to communicate with satellites
- for cooking (in microwave ovens)

Long wavelength

- broadcasting radio and TV programmes
- communicating with ships, aeroplanes and satellites

Now try this

1. Explain how X-rays are useful to doctors, but can also be a danger to patients. **(3 marks)**

2. Describe the similarities and differences between cooking using microwaves and cooking using infrared. **(3 marks)**

Ionising radiation

All electromagnetic waves transfer energy. The amount of energy transferred depends on the frequency of the wave. The higher the frequency, the more energy the wave transfers.

Remember:
higher frequency = more energy = more harmful.

Radioactive sources

Some elements are radioactive. These elements naturally emit ionising radiation all the time. This ionising radiation can be alpha (α) particles, beta (β) particles or gamma (γ) rays. All forms of ionising radiation transfer energy.

Effects of ionisation

The energy transferred by ionising radiation can remove electrons from atoms to form ions. Ions are very reactive and can cause mutations to the DNA in cells. This can lead to cancer.

Energy transferred by ionising radiation removes electrons from atoms to form ions.	→	Ions are reactive and can cause mutations to the DNA in cells.	→	Damaged DNA can lead to cancer.

Too much ionising radiation can kill cells, which is why gamma rays are useful for sterilising surgical instruments.

Worked example

Complete the table to show if different types of ionising radiation are particles or waves.

Radiation	Particles or waves?
alpha particles	particles
beta particles	particles
gamma rays	waves

It is important to remember which kinds of ionising radiations are particles and which are waves. Use this completed table to help you.

Now try this

target
D-B

1. Explain why gamma rays are ionising but visible light is not. **(3 marks)**

2. Explain why ionising radiation is dangerous. **(4 marks)**

Physics extended writing 1

To answer an extended writing question successfully you need to:

☑ Use your scientific knowledge to answer the question

☑ Organise your answer so that it is logical and well ordered

☑ Use full sentences in your writing and make sure that your spelling, punctuation and grammar are correct.

Worked example

William Herschel discovered infrared radiation in 1800, and Johann Ritter discovered ultraviolet radiation in 1801.

Describe the similarities and differences between the two discoveries. **(6 marks)**

Sample answer 1

Herschel measured the temperatures of different colours of light. Red was hottest for visible light. Infrared was even hotter. Ritter measured light and found that violet was the strongest. Ultraviolet was stronger. They are different because Ritter did his stuff after Herschel.

This is a basic answer. The statements about Ritter's experiment are too vague to score any marks, and 'stronger' is not a good way of describing Ritter's results.

Sample answer 2

Herschel split up sunlight using a prism, and investigated the temperature rise when thermometers were put in different parts of the visible spectrum. The temperature rise increased from the violet to the red end of the spectrum. He put a thermometer beyond the red end of the spectrum, and got an even greater temperature rise. This was caused by infrared radiation.

Ritter used silver chloride to detect the effects of the different parts of the visible spectrum. This turned black faster in violet light than it did in red. When he put the silver chloride just beyond the violet end of the spectrum it turned black even faster. This was due to ultraviolet radiation.

This is a good answer. It clearly explains the methods both men used.

Now try this

1. Ultraviolet radiation and X-rays are both parts of the electromagnetic spectrum. Compare their characteristics, uses and dangers. **(6 marks)**

The Universe

The Earth is one of eight planets that orbit around the Sun. The Sun is a star. Some of the planets have moons orbiting around them. The Sun, the planets and their moons all make up the Solar System.

Worked example

Explain the meanings of the words 'galaxy' and 'Universe'.

A galaxy is a collection of millions of stars. The Sun is part of a galaxy called the Milky Way. All the stars we can see in the sky at night are in the Milky Way.

There are billions of other galaxies. All the galaxies make up the Universe.

You need to remember the meanings of key words such as galaxy, Milky Way and Universe.

Watch out! A 'moon' (with no capital letter) means something orbiting around one of the other planets in the Solar System. The Moon (with a capital letter) is the object orbiting around the Earth.

Relative distances

In order of distance from Earth, starting with the closest:

Moon → Sun → planets in the Solar System → other stars → other galaxies

Earth   Moon

You could fit 30 'Earths' between the Earth and Moon.

Sun – over 11 000 'Earths' away →

Most of the planets are always further away from the Earth than the Sun. Mercury, Venus and Mars are sometimes closer.

Relative sizes

In order of size, starting with the smallest:

Moon → Earth → planets → the Sun and other stars → galaxies → Universe

You need to be able to compare the relative sizes of different things in the Universe, and also the distances between them.

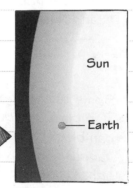

Sun

— Earth

The relative sizes of the Earth and the Sun

Now try this

target
D-B

1. Match up the diameters with the names of the objects.   **(2 marks)**
   Names: Earth, Jupiter, Milky Way, Moon, Sun.
   Diameters: 3500 km, 13 000 km, 143 000 km,
   1.4×10^6 km, 1.0×10^{18} km.

2. Suggest why people have visited the Moon, but only unmanned spacecraft have visited the other planets.   **(2 marks)**

Exploring the Universe

Modern telescopes

The first scientists explored the Universe by observing the visible light emitted by stars. They made more discoveries after telescopes were invented. Modern telescopes are very different to the early telescopes.

Development	Impact
greater magnifications	we can observe galaxies that are far away
recording observations using photography or digital cameras	we can gather more data
can be made with greater precision	we get clearer images
telescopes that can detect other parts of the electromagnetic spectrum	we can observe objects in space that emit more radio waves, infrared, ultraviolet or X-rays than visible light

Telescopes in space

Worked example

Some parts of the electromagnetic spectrum are absorbed by the Earth's atmosphere. Telescopes used to observe these wavelengths of radiation must be put on satellites in space. Use the graph to state which wavelengths must be observed by space telescopes.

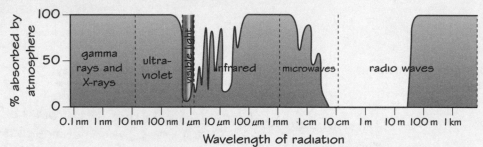

Telescopes must be in space to observe microwaves, infrared, ultraviolet, gamma rays and X-rays.

Some telescopes that use visible light are also placed in orbit. This is because clouds can block the view from Earth. Dust, water vapour and moving air can all distort visible light images, and in some places light pollution makes it difficult to see stars.

Investigating spectra

You can split the visible part of the electromagnetic spectrum into its different colours using a simple spectrometer made with a CD or DVD. Black lines in the spectrum show that some light is absorbed by the atmospheres of the Sun and Earth.

Now try this

1. Explain why an X-ray telescope must be put into orbit. **(2 marks)**

2. (a) Explain why visible light telescopes can be placed on the surface of the Earth. **(2 marks)**
   (b) Some visible light telescopes are put into orbit. Suggest two reasons for this. **(2 marks)**

Alien life?

The Earth is the only planet where we know that life exists. Scientists look for signs of life on other planets in several different ways.

Signs of life

 Life on Earth depends on liquid water. If there are signs that other planets have liquid water, then it is possible that life may exist there.

Plants on the Earth release oxygen during photosynthesis. Scientists can look for oxygen in the atmospheres of other planets, which may indicate life.

 Living organisms take in chemical substances as food and release other materials as waste. If one of these chemical changes is detected, it may be a sign of life.

EXAM ALERT!

If you are asked about signs of life in the exam remember that water only shows that life *may* exist on a planet or moon. It does not mean that life *does* exist.

Students have struggled with this topic in recent exams – **be prepared!**   ResultsPlus

Looking for the signs

Space probes can fly past planets or go into orbit around them. The probes can take images of other planets in the Solar System and send the information back to Earth.

Landers and rovers also investigate the surfaces of planets or moons. They can take close-up images, and test soil samples for microbes or for substances indicating life. A lander stays in one place on the planet, but a rover can move around.

This rover explored part of the surface of Mars.

Intelligent life?

If there is intelligent life on other planets, the 'aliens' may be trying to contact us using radio waves.

Watch out! Electromagnetic waves can travel through the vacuum of space. Sound waves *cannot* travel in a vacuum. Scientists are not looking for sound waves from alien civilisations. They are looking for radio waves.

Worked example

Explain what SETI is.

SETI stands for the Search for Extraterrestrial Intelligence. Information gathered by radio telescopes is analysed to see if there are any patterns in it, which could be a sign of intelligent life.

Now try this

 target D-B

1. Explain why scientists look for water and oxygen on other planets as possible signs of life. **(2 marks)**

2. Explain two things that a lander on Mars could do that cannot be done from an orbiter. **(4 marks)**

 target D-B

3. Explain how testing soil samples might show that life exists. **(2 marks)**

Life cycles of stars

The Sun formed about 4.5 billion years ago, and will last about another 5.5 billion years before it changes into a red giant. All stars that have a similar mass to the Sun go through the same stages in their life cycle.

The life cycle of a star is not like the life cycle of a living organism. Stars do not reproduce.

1 A nebula is a cloud of dust and gas. The gas is mostly hydrogen. The dust and gas can be pulled together by its own gravity. The hydrogen gets hotter as its gravitational potential energy is converted to kinetic energy (the particles move faster, so they are hotter).

It is important to remember that the force of gravity pulls the nebula together to make a star.

2 Eventually the gas gets hot enough for the nuclei of hydrogen atoms in the gas cloud to fuse together. Fusion reactions turn hydrogen atoms into helium and release a lot of energy. The star begins to shine. This stage of a star's life cycle is called the main sequence. The Sun is a main sequence star.

3 After about 10 billion years the star will have used up most of its hydrogen fuel. When this happens it swells up to become a red giant. Other elements fuse together in the collapsing core of the red giants.

4 Eventually all the elements that can take part in nuclear fusion reactions are used up. Fusion reactions in the star stop. Gravity pulls the material together to make a much smaller white dwarf star. The star gradually cools down.

Stars take billions of years to complete their life cycles, not millions.

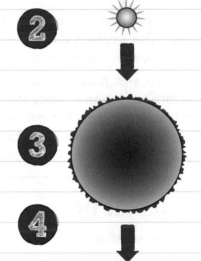

Higher-mass stars

A star with a much higher mass than the Sun follows the same first stages of the life cycle, but each stage is shorter. When most of its hydrogen is used up it forms a red supergiant. At the end of this stage the star will explode as a supernova. If what remains after the explosion is less than four times the mass of the Sun it will be pulled together by gravity to form a very small, dense star called a neutron star. More massive remnants form black holes.

Now try this

 target D-B

1. Explain why a white dwarf star begins to cool down. **(2 marks)**

 target C-A

2. Describe the role of gravity in the different stages of the life cycles of stars. **(2 marks)**

3. Explain why the Sun will not explode in a supernova. **(2 marks)**

Theories about the Universe

There are two different theories about the Universe.

> The Big Bang theory says that the whole Universe started out as a tiny particle about 13.5 billion years ago. The Universe expanded from this point in space. The Universe is still expanding today.

> The Steady State theory says that the Universe has always existed. It is expanding, and new matter is being created as it expands.

Red-shift

If a vehicle with a siren goes past you, you can hear the sound change. The movement of the siren changes the frequency and wavelength of the sound waves that you hear.

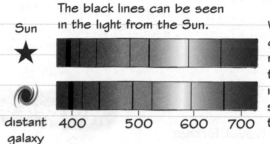

The black lines can be seen in the light from the Sun.

Sun

distant galaxy   400   500   600   700

When a distant galaxy is moving away, the black lines in its spectrum shift towards the red end.

A similar effect happens with light. Black lines in the spectrum of light from a star are moved closer to the red end of the spectrum if the star is moving away from us. This is called red-shift. The further the lines are shifted, the faster the object is moving away.

Light from distant galaxies shows red-shift. The more distant the galaxy the greater the red-shift. This shows that the Universe is expanding. Red-shift supports both theories, as both theories say that the Universe is expanding.

Worked example

Describe the cosmic microwave background radiation (CMB) and explain why the Big Bang theory is the currently accepted model for the beginning of the Universe.

You need to be able to describe the cosmic microwave background radiation.

The cosmic microwave background radiation is detected by radio telescopes and comes from all over the sky. The Big Bang theory says that this radiation was released at the beginning of the Universe.

Red-shift supports both the Big Bang and the Steady State theories, but the CMB only supports the Big Bang theory. The Big Bang theory is accepted because it has the most evidence supporting it.

Now try this

target B-A*

1. Light from a star moving away from the Earth is red-shifted. State how its wavelength has changed and how this affects its frequency.   **(3 marks)**

2. Distances in space are sometimes measured in light years. Galaxy NGC55 is 7.08 million light years away. The Pinwheel galaxy is 21 million light years away.
   Explain the similarities and differences you would expect in the light from these galaxies as a result of their movement.   **(4 marks)**

Infrasound

Humans can hear sounds with frequencies between 20 Hz and 20 000 Hz.

Other frequencies

Sounds with frequencies lower than 20 Hz are called infrasound. Sounds with frequencies higher than 20 000 Hz are called ultrasound (see opposite page).

Students often confuse infrasound and ultrasound waves with other kinds of waves. Infrasound and ultrasound are sound waves. They just have different frequencies and wavelengths from the sounds that humans can hear.

'Ultra' and 'infra' have the same meanings for electromagnetic waves. Ultraviolet light has higher frequencies than visible violet light, and infrared light has lower frequencies than visible red light.

Animals

Infrasound can travel further through air, water or the ground than sounds with higher frequencies. Some animals, such as whales and elephants, use infrasound for communication. They can make infrasound, and they can also hear it.

Worked example

Humans cannot hear infrasound. Explain how humans can use infrasound to study animals, and why this is important.

Scientists can use special instruments to detect sounds the animals are making. This allows scientists to follow the movements of the animals. Information about where animals are moving can help scientists to study them and to recommend how animals can be protected.

Meteors and volcanoes

Meteors burn up as they travel through the Earth's atmosphere. → As they pass through the air they produce infrasound waves. → Scientists can detect these waves. → They work out how many meteors enter the atmosphere and the paths they follow.

Erupting volcanoes produce infrasound. → The infrasound is detected by instruments all over the Earth. → Scientists detect volcanoes erupting in remote areas. → Governments can send help to people in the area.

Scientists learn more about volcanoes and how they erupt.

Now try this

1. Describe the similarities and differences between 'normal' sound, ultrasound and infrasound. **(3 marks)**

2. Suggest why whales use infrasound for communication as well as normal sounds. **(2 marks)**

3. Explain why it is useful to detect eruptions using remote sensing (such as listening for infrasound). **(3 marks)**

Ultrasound

Animals

Animals such as bats and dolphins can make and detect ultrasounds. They use these sounds to communicate with one another.

Scans

Ultrasound waves are used to make images of the inside of the body. Ultrasounds are not harmful, so it is safe to use them to scan foetuses (unborn babies). The ultrasound waves are sent into the woman's body, and some of the sound is reflected each time it meets a layer of tissue with a different density to the one it has just pased through. The scanner detects the echoes and a computer uses the information to make a picture.

Sonar

Sonar uses pulses of ultrasound to find the depth of water beneath a ship. The sonar equipment measures the time between sending the sound and detecting its echo. This time is used to calculate the depth of the water, using this equation:

distance = speed × time
(see page 71)

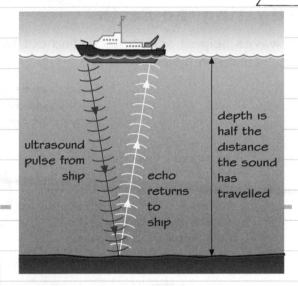

ultrasound pulse from ship

echo returns to ship

depth is half the distance the sound has travelled

Worked example

A ship detects an echo 5 seconds after it has sent out a sonar pulse. Sound travels at 1500 m/s in sea water. How deep is the water?

$$\text{distance} = \text{speed} \times \text{time}$$
$$= 1500 \text{ m/s} \times 5 \text{ sec}$$
$$= 7500 \text{ m}$$
$$\text{Depth of water} = \frac{7500 \text{ m}}{2}$$
$$= 3750 \text{ m}$$

Be careful with the units. If the speed is in metres per second, the time must be in seconds.

Remember that the depth of the water is *half* the distance the ultrasound has travelled.

Now try this

target
C-B

1. Describe how an ultrasound scan produces a picture of a foetus.
   (4 marks)

2. A sonar signal is sent out from a boat. The echo is received 0.2 s later. The sound has travelled a total distance of 280 m. Calculate the speed of sound in water.
   (3 marks)

Seismic waves

P waves and S waves

Seismic waves can be caused by earthquakes or explosions. Seismic waves can be P waves or S waves. P waves are longitudinal waves, and S waves are transverse waves. P waves travel faster than S waves. All seismic waves can be detected using seismometers.

You can remember the different types of wave using this method:

P waves push and pull the rock as they pass, so they are longitudinal.

S waves move the rocks side to side as they pass, so they are transverse.

Investigating the Earth

Seismic waves helped scientists to find out that the Earth is made of different materials in the core, mantle and crust.

Seismic waves from earthquakes travel through the Earth.

Part of the wave is reflected at the boundaries between different layers.

The speed of seismic waves changes gradually with depth because the rocks have different properties. The gradual change in wave speed makes waves travel along a curved path.

- crust
- mantle
- core

Worked example

Explain how seismic waves can be used to find out where an earthquake happened.

P waves are primary waves, so they arrive first. S waves are secondary waves, so they arrive second.

Seismic waves spread out in all directions from the point where the earthquake happens, and the waves are detected by seismometers. P waves travel faster than S waves, so they are detected first. The greater the distance between the earthquake and the seismometer, the greater the difference in arrival times. Scientists use the difference in the arrival times of the P waves and S waves to work out the distance from each seismometer to the place where the earthquake happened. Distances from at least three seismometers in different places are used to work out where the earthquake happened.

Now try this

1. Describe the similarities and differences between S waves and P waves. **(3 marks)**

2. Explain how the position of an earthquake is located using P and S waves. **(4 marks)**

3. Explain why seismic waves travel on curved paths within the Earth. **(3 marks)**

Predicting earthquakes

Earthquakes cause a lot of damage, and people can be killed by falling buildings. If an earthquake happens beneath the sea it can cause a huge wave called a tsunami. Tsunamis kill thousands of people and destroy buildings near the coast.

Scientists can give people warnings about tsunamis. Seismometers are used to find the location of an earthquake. If the earthquake happened beneath the sea then they can warn people in nearby countries that a tsunami may be coming. People living near the coasts could have several hours warning.

Tectonic plates

The outermost layer of the Earth is divided into sections called tectonic plates. The plates move slowly, driven by convection currents in the Earth's mantle.

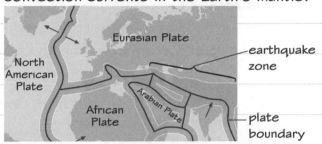

You don't need to know the names of different tectonic plates or their boundaries for your exam.

How earthquakes happen

Tectonic plates move slowly driven by convection currents in mantle ➤ No gaps between plates so their edges slide past each other ➤ Friction between plates means they don't slide smoothly ➤ Forces build up until part of the plate breaks ➤ Plates move with a sudden jerk

⟶ EARTHQUAKE

Worked example

Give two reasons why scientists cannot predict when an earthquake will happen.

1. They don't know how big the forces acting on the tectonic plates are.
2. They don't know the amount of friction on the plates or how strong they are so they can't predict when they will break.

The question asks for two reasons. It is a good idea to number your answer to make sure that you have written down two different reasons.

Where do earthquakes happen?

Earthquakes happen at the boundaries between tectonic plates. This means scientists can predict where an earthquake might take place, even if they can't predict when.

Now try this

target **C-B**

1. A tsunami in Japan in 2011 was caused by an earthquake 70 km from the coast. The tsunami waves travelled at around 45 m/s.
   (a) What is the maximum possible warning people on the closest part of the shore could have had? **(4 marks)**
   (b) Suggest one reason why the warning time would have been less than this. **(1 mark)**

2. Explain why scientists cannot predict when an earthquake will happen. **(2 marks)**

Your exam paper will include a page with all the equations you may need to answer the questions. For this question, look at page 71.

Physics extended writing 2

Worked example

The Sun has a mass of 2×10^{30} kg. Rigel is a star with a mass of about 48×10^{30} kg. Compare the life cycles of these two stars. **(6 marks)**

Sample answer 1

The Sun will become a red giant star, and then it will shrink to become a white dwarf. Rigel will become a red supergiant and it will then explode in a supernova.

This is a basic answer. This student has named most of the key stages of the two life cycles correctly, and also has the sequence correct. However they have not said what happens to a high mass star after a supernova, and there is not enough detail about the different stages.

Sample answer 2

The first part of the life cycle is similar for both stars. They both formed when gravity pulled the dust and gas in a nebula together. The gas heated up as it was compressed, and eventually fusion reactions started that turned hydrogen into helium. The outward pressure from the energy released by fusion stopped the star collapsing any more. This is the main sequence stage of the star. This stage will be much shorter for Rigel because it has a higher mass.

When most of the hydrogen is used up, both stars will form giants. The Sun will form a red giant, and Rigel will form a much bigger red supergiant. Their life cycles are different after this.

When the red giant phase of the Sun is over it will throw off a shell of gas and what is left will be pulled together by gravity to form a small white dwarf which will gradually cool down. At the end of its red supergiant phase, Rigel will explode. This is called a supernova. What is left will form a neutron star or a black hole.

This is an excellent answer. The student has given a lot of details about the various stages in the life cycles, and has also clearly described where the two life cycles are similar and where they are different. The answer is organised into paragraphs.

Now try this

1. Today's astronomers use telescopes that make images using the whole range of wavelengths in the electromagnetic spectrum. Telescopes on the surface of the Earth can be larger, cheaper and easier to repair than telescopes on satellites.

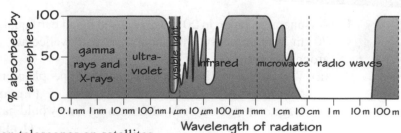

Explain why some telescopes are put onto satellites.
Use information from the graph to illustrate your answer.

(6 marks)

Physics extended writing 3

Worked example

Planets orbiting around other stars in our galaxy can sometimes be detected because they make the star 'wobble' as they move around their orbit. This wobble can be detected using a technique called Doppler spectroscopy, which detects very small changes in the speed of the star moving towards or away from us.

Explain what you would see if the planet is making the star wobble and how the changes in the light coming from such a star are different to the red-shift that is seen in the light from distant galaxies. Your answer should make it clear what red-shift is.   **(6 marks)**

Sample answer 1

Red-shift is the change in frequency of light coming from distant galaxies. Red-shift is evidence for the Big Bang theory and the Steady State theories, because both these theories say that the Universe is expanding. The Big Bang theory is the one that astronomers accept because it is also supported by evidence of the Cosmic Microwave Background radiation.

This is a basic answer. There is a lot of correct science here, but unfortunately the only part that answers the question is the statement that implies that the red-shift in light from galaxies is caused by the Universe expanding. It doesn't matter how much good science you write – you only get marks if you answer the question on the paper!

Sample answer 2

Red-shift is a change in the frequency of light from the stars caused by the object emitting the light moving away from us. This change is detected because black lines in the spectrum of light from a moving star are shifted towards longer wavelengths. If the object is moving towards us the light will be shifted to shorter wavelengths.

The lines from a star with a planet orbiting it will move first one way and then the other. This is different to light from distant galaxies, which always show red-shift. If the galaxy is further away the red-shift is greater.

The changes in the light from the star with a planet is due to the star wobbling. The red-shift of galaxies is due to the Universe expanding.

This is an excellent answer. It contains all the necessary details, and is presented in a sensible order.

Now try this

1. On 11 March 2011 there was a large earthquake beneath the sea 70 km west of Japan. This earthquake caused a tsunami that killed about 15,000 people. Many more people would have been killed if there had not been warnings that a tsunami was on the way.

   Using your knowledge of earthquakes and tsunamis, explain why it is hard to predict an earthquake but it is sometimes possible to give up to 5 hours warning of a tsunami approaching.   **(6 marks)**

Renewable resources

Electricity is not a source of energy. It has to be generated using renewable or non-renewable energy resources.

1 Renewable energy resources are resources that are easily replaced.

For example, solar energy can be used in special power stations to generate electricity. Solar cells can also turn energy from the Sun into electricity directly.

2 The wind can be used to turn turbines, and the turbines power generators. Waves can generate electricity in floating generators or in small power stations built on the coast.

These resources depend on the weather, so they are not available all the time. Solar energy is more useful in countries sunnier than the UK!

3 Hydroelectricity is generated by flowing rivers or water falling from a dam in the hills.

4 Geothermal energy uses heat underground to turn water to steam, and the steam drives turbines.

5 Biomass can also be used to generate electricity.

These renewable resources are available all the time. However, there are not many suitable places to use these in the UK.

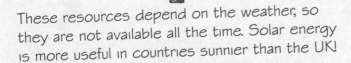 Worked example

Explain how electricity can be made using energy in the tides.

A tidal barrage can be built across an estuary to trap water as the tide goes out. The trapped water then flows through turbines to make them spin. The turbines turn a generator that generates electricity. Turbines can also be put under the sea where tidal currents flow.

You need to remember key words such as **tidal barrage** (a dam built across a river or estuary) and **turbine** (a machine that spins when wind or water flows past it).

Advantages and disadvantages

☑ Generating electricity using renewable resources does not usually cause pollution.

☑ There are no fuel costs.

☒ It does cost money to build the machinery to generate electricity.

☒ Many people think that wind turbines spoil the view.

☒ The reservoirs used for hydroelectricity flood valleys, and tidal barrages change the environment in river estuaries. Both of these change habitats and can harm wildlife.

☒ Most forms of renewable energy are not available all the time.

 Now try this

1. Compare the availability of tidal power and hydroelectricity. **(3 marks)**

2. Compare the advantages and disadvantages of wind power and geothermal energy. **(4 marks)**

Non-renewable resources

Non-renewable resources include fossil fuels (coal, oil and natural gas) and nuclear fuel. Most electricity in the UK is generated from non-renewable resources.

Advantages and disadvantages of fossil fuels

✓ At present, generating electricity using fossil fuels is cheaper than using renewable resources.

✓ Electricity from power stations is available all the time (unlike electricity from renewable resources).

> You need to remember the polluting gases produced by burning fossil fuels.

✗ All non-renewable resources will eventually run out and cannot easily be replaced.

✗ Burning fossil fuels releases carbon dioxide (which is contributing to climate change) and sulfur dioxide (which causes acid rain).

✗ Extracting and transporting fossil fuels causes pollution and can change landscapes.

Nuclear power

✓ Nuclear power stations do not release carbon dioxide or other polluting gases.

✗ The uranium used as nuclear fuel will run out one day.

✗ Many people are worried about the possibilities of an accident in a nuclear power station, and do not want to use nuclear energy.

✗ An accident in a nuclear power station could affect thousands or even millions of people. Nuclear power stations must be safe, making them much more expensive to build than fossil-fuelled power stations. Nuclear power stations are also very expensive to decommission (dismantle) at the end of their life.

✗ Nuclear power stations produce radioactive waste, which stays radioactive for millions of years. This must be sealed into glass or concrete and buried so that the radioactivity cannot damage the environment.

Now try this

target
D-B

1. Explain one advantage that nuclear power stations have over fossil-fuelled power stations, and one disadvantage. **(2 marks)**

2. Compare the advantages and disadvantages of using fossil fuels instead of renewable resources to generate electricity. **(4 marks)**

EXAM ALERT!

Give details if you are answering a question about fossil fuels. Don't just say 'causes pollution', but say which gases are released.

Results Plus

Students have struggled with exam questions similar to this – **be prepared!**

Generating electricity

Inducing a current

If you move part of a loop of wire in a magnetic field, an electric current will flow in the wire. This is called electromagnetic induction, and the current is an induced current.

You can get the same effect by keeping the wire still and moving the magnet.

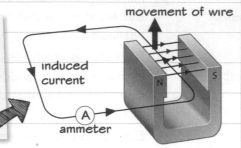

movement of wire

induced current

N   S

(A) ammeter

You can change the direction of the current by:
- changing the direction of motion of the wire
- changing the direction of the magnetic field.

You can increase the size of the current by:
- moving the wire faster
- using stronger magnets
- using more loops of wire, so there is more wire moving through the magnetic field.

These factors can be investigated in the laboratory using a magnet, a coil of wire and a sensitive voltmeter or ammeter.

Generators

This diagram shows a simple generator. The generators in power stations work in a similar way to the one in the diagram. However, they need to use very strong magnetic fields, so they usually use electromagnets instead of permanent magnets.

Permanent magnets produce a magnetic field. The stronger the magnetic field, the greater the current.

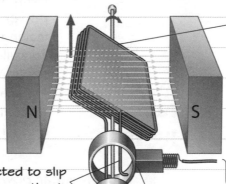

A coil is wound on an iron core. The many turns of wire and the iron core increase the size of the current.

N   S

The ends of the coil are connected to slip rings. These allow the coil to spin without twisting the wires to the rest of the circuit.

Induced alternating current (a.c.)

Carbon brushes press on the slip rings to make electrical contact with the rest of the circuit.

An alternating current changes direction many times each second. Generators produce alternating current because each side of the coil goes up through the magnetic field and then comes down again. This induces a current in one direction then in the opposite direction.

Direct current (produced by batteries) always flows in the same direction.

Now try this

target **D-C**

1. Suggest three ways in which the current produced by the generator could be increased. **(3 marks)**

2. Explain why generators in power stations use electromagnets instead of permanent magnets. **(2 marks)**

target **C-A**

3. Copy the graph axes shown and draw labelled lines on it to show:

   current | O | + | − | time

   (a) direct current **(1 mark)**

   (b) alternating current. **(1 mark)**

Transmitting electricity

The National Grid

The National Grid is the system of wires that transmit electricity from power stations to the places where the electricity is used. Electricity is transmitted at high voltages. Increasing the voltage reduces the current. This improves the efficiency because less energy is wasted as heat in the transmission lines.

Transformers are used to change the size of an alternating voltage. Step-up transformers increase the voltage as it leaves the power station and before the electricity is transmitted. The voltage is high enough to kill you if you touch a transmission line, or if a kite string or carbon fishing pole touches the line. Step-down transformers take electricity from the power lines and reduce it to safer voltages again before it goes into homes.

EXAM ALERT!

Remember that increasing the voltage using a transformer *decreases* the current. If the current is lower the thermal energy losses are smaller, so transmission at high voltages is more efficient.

Students have struggled with this topic in recent exams – **be prepared!**   ResultsPlus

Transformer calculations

The voltage produced by a transformer depends on the ratio of the number of turns in the primary and secondary coils, and on the voltage in the primary coil.

$$\frac{\text{voltage (primary)}}{\text{voltage (secondary)}} = \frac{\text{turns (primary)}}{\text{turns (secondary)}}$$

$$\frac{V_P}{V_S} = \frac{N_P}{N_S}$$

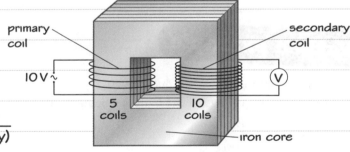

For the transformer in the diagram:

$$\text{voltage (secondary)} = \text{voltage (primary)} \times \frac{\text{turns (secondary)}}{\text{turns (primary)}}$$

$$= \frac{10\,V \times 10}{5}$$

$$= 20\,V$$

> If the number of turns on the secondary coil is greater than the number of turns on the primary coil, it is a step-up transformer.

Now try this

target E-C

1. Explain why electricity is sent around the National Grid at high voltages. **(2 marks)**

target C-A

2. A transformer has 50 turns on the primary coil and 1000 turns on the secondary coil. If the secondary voltage is 400 kV, calculate the primary voltage. **(2 marks)**

Electrical power

Current and voltage

Electricity is a flow of charged particles. Current is the amount of charge flowing past a point each second. Voltage is like an electrical pressure – it measures the amount of energy transferred by a current.

Power

The power of an appliance is the amount of energy it transfers every second. Energy is measured in joules (J) and power is measured in watts (W).

EXAM ALERT!

Power is *how fast* energy is being transferred. Remember that 1 watt is 1 joule being transferred every second.

Students have struggled with this topic in recent exams – **be prepared!**   ResultsPlus

---

Calculating power

Power can be calculated using this equation:

$$P = \frac{E}{t}$$

P = power (watt, W)
E = energy (joule, J)
t = time (second, s)

The power of an electrical appliance can also be calculated using the current and the voltage:

$$P = I \times V$$

P = power (watt, W)
I = current (amp, A)
V = voltage (volt, V)

Worked example

The current in a 1.2 W torch bulb is 0.2 A. Calculate the voltage of the batteries being used.

$$\text{voltage} = \frac{\text{power}}{\text{current}}$$

$$= \frac{1.2\,W}{0.2\,A}$$

$$= 6\,V$$

You do not need to memorise the equations, as they will be given to you in an exam. But you *do* need to be able to choose the correct equation to use, and to rearrange it if necessary (see page 102). And don't forget the units!

Finding the power

This circuit can be used to find the power of an appliance.

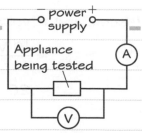

Now try this

target **C-B**

1. A 60 W light bulb transfers 1200 J of energy. Calculate how long it was switched on. **(3 marks)**

2. A kettle with a power of 1000 W is plugged into the 230 V mains supply. Calculate the current in the kettle. Give the unit. **(4 marks)**

3. An electric heater with a power of 2 kW is switched on for one minute. Calculate the amount of energy it transfers. **(4 marks)**

Paying for electricity

Power ratings

The energy transferred by an appliance depends on its power and the time it is switched on for. The power rating of an appliance tells you how much energy it uses each second.

Some appliances transfer a lot of energy. Their power ratings are often given in kilowatts.

1 kilowatt (kW) = 1000 W

Units for energy

In most science work we use the joule as the unit for energy. Electricity companies use a much larger unit for energy, the kilowatt-hour (kWh). This is the amount of energy transferred by a 1 kW appliance in 1 hour.

> Don't get kilowatts and kilowatt-hours mixed up! Remember that kilowatts are a unit for power, and kilowatt-hours are a unit for energy.

Worked example

It takes 3 minutes to boil a 2 kW kettle. Electricity costs 15 p/kWh. How much does it cost to boil the kettle?

cost (p) = power (kilowatt, kW) × time (hour, h)
         × cost of 1 kW h (p/kWh)

The time must be in hours.

$3 \text{ minutes} = \dfrac{3}{60} \text{ hours}$

            = 0.05 hours

cost   = 2 kW × 0.05 h × 15 p/kWh

         = 1.5 p

> Be careful with your units! If you are using the equations for calculating power on the opposite page then the power must be in watts and the time in seconds.
>
> If you are calculating the cost of electricity, the power must be in kilowatts and the time in hours.

Check your answer to see if it is sensible:
- An electricity bill for 3 months shouldn't be more than around £200 (or 20 000p).
- The cost of running an appliance for a few hours shouldn't be more than a few pounds.
- The cost of running an appliance for a few minutes shouldn't be more than a few pence.

If you get a very large number, check that you have used the correct units – the power should be in kW and the time should be in hours.

Now try this

target
B-A*

1. A 100 watt (0.1 kW) bulb is switched on for 6 hours. Electricity costs 14p per unit.
Calculate the cost. **(2 marks)**

2. A 2 kW electric fire is switched on for 2 hours. It costs 50p to run it.
Calculate the cost of 1 kW h. **(2 marks)**

Reducing energy use

Using less energy saves us money. It also helps the environment because most forms of energy we use to generate electricity result in carbon dioxide being added to the atmosphere.

Using less energy

Use efficient light bulbs   **Use microwave ovens**   **Insulate homes**

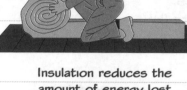

Old light bulb:
10 W of light output
100 W of energy input

Modern light bulb:
10 W of light output
20 W of energy input

- use less energy than a normal oven
- cook food more quickly
- but some people don't like them because the food does not go brown.

Insulation reduces the amount of energy lost from homes, and reduces heating bills.

Payback times

Although you will save some money by using a low-energy appliance or insulating your home, you need to spend money to buy the appliance or the insulation. The payback time is the time it takes to save the amount of money you had to spend to start with.

$$\text{payback time} = \frac{\text{cost of energy-saving method}}{\text{savings per year}}$$

Worked example

It costs a homeowner £75 to have extra insulation added to their loft. They should save about £25 per year on their energy bills. What is the payback time?

$$\text{payback time} = \frac{£75}{£25 \text{ per year}}$$

$$= 3 \text{ years}$$

Making decisions

The most cost-efficient method is the one with the shortest payback time. This is the one that will give you the biggest savings for each pound you spend.

Cost isn't always the only thing to consider when buying new appliances. For example, you might want a new fridge because your current one is too small.

Now try this

target
D-B

1. It costs £2000 to fit double glazing to a small house. The savings per year are £150. What is the payback time? **(2 marks)**

2. An energy-efficient washing machine costs £100 more than a less efficient one. It costs 10p less to run for each wash.

   Explain what other information you need to calculate the payback time for buying the more expensive machine. **(2 marks)**

Energy transfers

Energy is never created or destroyed. The total energy before an energy transfer is exactly the same as the total energy afterwards. The energy has just been changed into one or more different forms. This is the law of conservation of energy.

Forms of energy

Energy can be changed from one form to another. The forms of energy are:
- thermal (heat energy)
- light
- sound
- electrical
- kinetic (movement)
- chemical (energy stored in food, fuel, etc.)
- nuclear (energy stored within atoms)
- elastic potential (energy stored in stretched springs, etc.)
- gravitational potential (energy stored in objects in high positions)

Energy transfer chains

This is an energy transfer chain for a battery operated radio. It shows the different forms of energy in the radio.

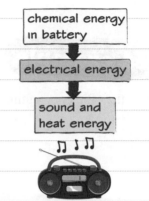

chemical energy in battery
↓
electrical energy
↓
sound and heat energy

Worked example

The diagram shows the energy transfers in a kettle.
(a) Fill in the missing value.
(b) Calculate how much energy is wasted when the kettle is boiled.

(a)

Heat energy to kettle and surroundings
28 kJ

Electrical energy 210 kJ

Sound energy 2 kJ

Heat energy to water 180 kJ

(b) 28 kJ + 2 kJ = 30 kJ

In an energy transfer diagram the width of each arrow represents the amount of each type of energy. You may be asked to fill in missing numbers on diagrams like this, or to explain what they show.

The total amount of useful and wasted energy transferred is the same as the energy that was supplied to the kettle.
$210 - 180 - 2 = 28$

The heat energy transferred to the water in the kettle is **useful energy**. The rest of the energy is **wasted energy**.

Now try this

target D-B

1. A kettle uses 375 kJ of electrical energy to heat some water. 300 kJ of energy ends up as useful thermal (heat) energy in the water. Calculate the amount of energy wasted.   **(2 marks)**

target C-A

2. (a) Draw a flow chart to show the energy changes when a box falls off a high shelf onto the floor.   **(3 marks)**
   (b) Explain how your flow chart illustrates the law of conservation of energy.   **(2 marks)**

Efficiency

All machines waste some of the energy they transfer. Most machines waste energy as heat energy. The efficiency of a machine is a way of saying how good it is at transferring energy into useful forms.

A very efficient machine has an efficiency that is nearly 100%. The higher the efficiency, the better the machine is at transferring energy to useful forms.

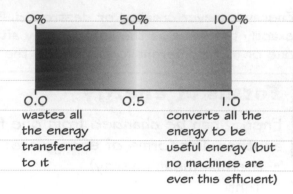

0%	50%	100%
0.0	0.5	1.0
wastes all the energy transferred to it		converts all the energy to be useful energy (but no machines are ever this efficient)

Worked example

What is the efficiency of this television?

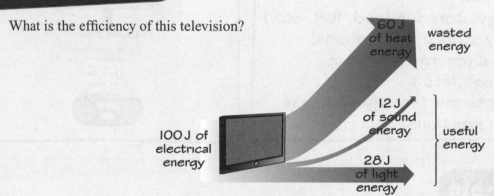

100 J of electrical energy

60 J of heat energy — wasted energy

12 J of sound energy ⎫
28 J of light energy ⎬ useful energy

$$\text{efficiency} = \frac{\text{useful energy transferred by the machine}}{\text{total energy supplied to the machine}} \times 100\%$$

◀ Efficiency does not have units.

Total useful energy = 12 J + 28 J
= 40 J

$$\text{efficiency} = \frac{40\,J}{100\,J} \times 100\%$$
= 40%

◀ No machine is ever 100% efficient. If you calculate an efficiency greater than 100% you have done something wrong!

Now try this

1. An old-style filament lamp uses 60 J of electrical energy, and transfers 6 J of this to light energy.

   (a) How much energy is wasted?   **(1 mark)**

   (b) What is the efficiency of the filament lamp?
                                    (2 marks)

60 J of electrical energy each second

6 J of light energy each second

2. A light emitting diode (LED) has an efficiency of 80%.
It emits 8 J of light energy every second.
Calculate how much electrical energy is supplied to it each second.   **(3 marks)**

The Earth's temperature

Energy balance

Any object that has a constant temperature must be absorbing the same amount of power as it is radiating.

The warm water loses heat to its surroundings.

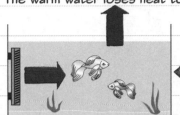

power in = power out
temperature constant

The water is heated by an electric heater.

If the power absorbed is greater than the power radiated, it will warm up.

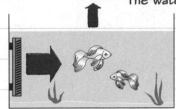

The heater is switched to a higher power setting. Power in is greater than power out → temperature rises

If the power absorbed is less than the power radiated, it will cool down.

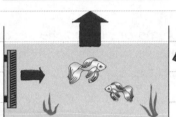

The heater is switched to a lower power setting. Power out is greater than power in → temperature falls

Remember that power is the amount of energy transferred each second.

Worked example

Peter investigates the factors that affect the amount of radiation absorbed by different coloured objects. He uses cans of water painted black, grey and white, and stands them in the sun. Explain which can will have the hottest water after 5 minutes.

The water in the black can will be the hottest, because black surfaces absorb more thermal energy than paler surfaces.

Black surfaces absorb more thermal energy than white surfaces. Black surfaces also radiate more thermal energy than white ones.

The Earth

The Earth is warmed by absorbing energy radiated by the Sun. The warm Earth radiates radiation into space.

If the power absorbed from the Sun is the same as the power radiated, the Earth stays at the same temperature.

Some gases in the atmosphere act as 'greenhouse gases'. They absorb some energy radiating from the Earth and stop it being radiated away into space. This means that the amount of radiation emitted by the Earth is now less than the amount absorbed. The Earth warms up.

Now try this

target D-C

1. Look at the diagrams of the fish tank. Explain what will happen to the temperature of the water in this tank if you put a lid on the tank. **(3 marks)**

target C-A

2. Snow often takes a long time to melt, even on sunny days. Snow melts faster if you spread soot on it. Explain these statements. **(3 marks)**

Physics extended writing 4

Worked example

Transformers are used to change the voltage and current of an alternating electricity supply. Asif is investigating the efficiency of transformers with different numbers of turns of wire on the secondary coil. His hypothesis is that the transformer will be more efficient when the secondary voltage is higher.

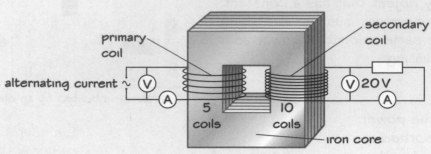

Describe how Asif can test his hypothesis.

(6 marks)

Sample answer 1

Test the transformer with different numbers of turns of wire on the secondary coil to give different secondary voltages. Calculate the power in and the power out. Take the secondary power and divide it by the primary power to find the efficiency for each different coil.

This is a basic answer. This student has got the right idea, but has not explained in enough detail. If you are asked to describe an investigation you need to give a lot more detail, such as the number of different conditions you will try, how you will make your test fair, how you will process your results and how you will decide whether or not the hypothesis is correct.

Sample answer 2

Measure the current and voltage in the primary and secondary circuits, then change the number of turns of wire on the secondary coil. Measure the currents and voltages again. Do this for 5 different numbers of turns on the secondary coil.

Calculate the power in the primary coil each time by multiplying the current and voltage. Do the same to calculate the power in the output coil. Calculate the efficiency for each different secondary coil by dividing the power out by the power in and then multiply by 100 to give a percentage.

Keep the test fair by using the same iron core in the transformer for all the tests, and the same kind of wire and the same primary circuit.

This is a good answer. The student could improve it by saying more about how to process the results to test the hypothesis. For example, the efficiency could be plotted against output voltage on a graph to make it easier to see if there is a relationship between the secondary voltage and the efficiency. The hypothesis would be correct if this graph gave a line sloping upwards from left to right.

Now try this

1. Discuss the advantages and disadvantages of the use of tidal barrages and hydroelectricity for generating electricity in the UK.

(6 marks)

Physics extended writing 5

If the polar ice caps melt, dark seas will be exposed in the Arctic and dark rocks in the Antarctic. Cutting down conifer forests in snowy areas allows more white snow to be visible.

Explain how the Earth maintains a constant average temperature, and how the changes above could affect its temperature. **(6 marks)**

Sample answer 1

Power emitted = power absorbed = constant temperature.

Less ice = less reflected = warmer.

More snow = more reflected = cooler.

This is a basic answer. The student appears to understand the science involved, but has not really explained it very well. The answer here would be good as a *plan* for a proper answer, but answers need to be written in full sentences and provide full scientific explanations.

Sample answer 2

The Earth absorbs energy from the Sun, and also radiates energy back into space. As long as the power radiated is the same as the power absorbed the Earth will maintain a constant temperature.

The polar ice caps are white, so they reflect most of the radiation that falls on them back into space. If they melt, the darker surface of the water or rock will absorb more radiation. The Earth will now be absorbing more power than it is radiating and it will warm up. As it gets warmer it radiates more energy. The temperature will become steady when the power radiated is equal to the higher amount being absorbed. The Earth's steady temperature will be higher than before.

This is a good answer. It explains step by step the effect that a change in polar ice caps will have, and also states how the new steady temperature will be different to the original. It also mentions power, which is better than writing about the 'energy' or 'heat' being absorbed or radiated by the Earth. To make it an excellent answer the student should also write about the effect of cutting down conifer forests.

1. Usma is investigating how the power supplied to an appliance changes when the voltage is changed. Her hypothesis is that the power increases when the voltage from the supply is increased. She uses the circuit shown.
    Explain how Usma can use this apparatus to test her hypothesis. **(6 marks)**

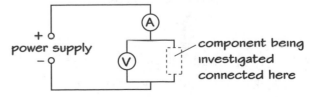

Practical work 1

Questions on practical work

The Edexcel Science course includes suggestions for a lot of different investigations. By the time you sit your exams you will have completed a Controlled Assessment, based on one or more of these investigations. But you could also be asked questions about any of these practicals in the exam.

Questions based on practical work could include:
- writing a method for an investigation
- explaining how to carry out a fair test
- drawing a graph to show some results
- writing a conclusion based on results given in the exam paper
- evaluating a method or a conclusion.

Worked example

A student is investigating the effectiveness of different indigestion remedies. He is using a pH meter to find out how the pH of hydrochloric acid changes when he adds one dose of each indigestion remedy.

Explain how the student can control three different variables so that the results can be compared.

(1) He should use the same volume of acid each time, by measuring it into the beaker using a measuring cylinder.

(2) He should use the same concentration of acid each time. He can do this by using acid from the same bottle each time.

(3) He should leave each indigestion remedy in the acid for the same length of time before taking the pH reading. He can use a stopclock to tell him when to measure the pH.

This is a good answer, because the student has explained *how* to control each variable.

Other variables that could have been mentioned in this answer include keeping the temperature the same (as temperature can affect the rate of reactions), and stirring the mixture in the same way each time.

Other questions on planning

Other questions on the planning part of a practical could include asking you to:
- describe a method
- explain the apparatus needed.

Say why you need each piece of apparatus. Remember to include apparatus you may need to control variables.

You must describe a method in the correct order. It may help to jot down some ideas in a blank space on your exam paper to help you to get your ideas in order.

If the question asks you to explain the method, remember to say *why* each step is needed.

Practical work 2

Dealing with evidence

Worked example

The graph shows the results of an investigation to find out how the current flowing through an appliance changes when the voltage is changed. Complete the graph by drawing a line of best fit.

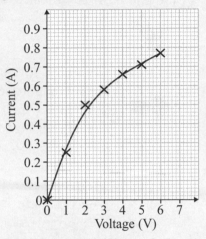

A result that does not fit the pattern is an **anomalous result**. The student has probably made a mistake when measuring or recording this result.

Do not include anomalous results when you are working out means, or when you are drawing lines or curves of best fit.

EXAM ALERT!

This is a good answer, because the student has drawn a smooth curve through most of the points. Do not join the points using straight lines. Try to include as many points as possible on your curve, but ignore any that are obviously not following the pattern of most of your results.

Students have struggled with exam questions similar to this – **be prepared!**

ResultsPlus

- -

Conclusions and evaluations

Worked example

A student investigating current and voltage in an appliance had the following hypothesis:

'The current will increase when the voltage increases'.

Look at the graph of their results (above).
(a) Write a conclusion for this investigation.
(b) Evaluate the quality of the conclusion.

(a) The graph shows that the current increases when the voltage increases. The hypothesis was correct.

(b) The points all lie close to a smooth curve, so there were probably not many errors in the measurements. The quality of the data could be improved by taking several measurements at each voltage and finding means (averages) of the results.

Only one appliance was tested, so the conclusion really only applies to this appliance. Several different appliances would need to be tested before we could test the hypothesis properly and know whether this conclusion applies to all appliances.

This would be a better answer if the conclusion described the shape of the graph in more detail. The graph is a curve, which shows that the increase in current for each step in voltage gets less as the voltage gets higher.

Stating whether or not the hypothesis is correct is good. But this would be better if the answer pointed out that the current is not proportional to the voltage in this appliance. Proportional means that the current doubles if the voltage doubles.

Taking at least three readings for each point and finding a mean is almost always a good way of improving the quality of data. To calculate a mean, add up all the results for each point and then divide by the number of results.

101

Final comments

Here are some other things to remember in your exam.

Read the question carefully. Underline important words in the question to help you to understand what you need to do. Using correct science is great, but no use if it does not actually answer the question!

Use the correct scientific words for things and don't be vague in your answers. For example, saying 'fossil fuels cause pollution' isn't specific enough. Saying *how* they cause pollution is better ('burning fossil fuels causes pollution because carbon dioxide and sulfur dioxide get into the air').

Know what the command words mean at the start of a question. If you are asked to 'explain' then you need to say what happens and how or why it happens. If a question asks you to 'compare' then you need to write down something about all the things you are comparing and how they are similar and different.

Don't use a formula for chemical substances unless the question asks for it. For example, if your answer is carbon dioxide and you write C_2O by mistake (because carbon dioxide is CO_2), it won't be possible to tell that you knew the right answer.

Learn how to balance chemical equations. And remember that all the gases that take part in the reactions you need to know about for this course are diatomic (their formulae are: O_2, Cl_2, N_2, and so on).

Revise the investigations you have carried out during the course. You may be asked questions on practical work in the exam.

Show your working for calculation questions, even if you use a calculator. And don't forget to include the units with your final answer.

Write your name here
Surname
Other names

Edexcel GCSE
Centre Number   Candidate Number

Chemistry/Science
Unit C1: Chemistry in Our World

Higher Tier

Time: 1 hour
Paper Reference
XXX

You must have:
Calculator, ruler
Total Marks

Instructions
- Use **black** ink or ball-point pen.
- **Fill in the boxes** at the top of this page with your name, centre number and candidate number.
- Answer **all** questions.
- Answer the questions in the spaces provided – there may be more space than you need.

Information
- The total mark for this paper is 60.
- The marks for **each** question are shown in brackets – use this as a guide as to how much time to spend on each question.
- Questions labelled with an **asterisk** (*) are ones where the quality of your written communication will be assessed – you should take particular care with your spelling, punctuation and grammar, as well as the clarity of expression, on these questions.

Advice
- Read each question carefully before you start to answer it.
- Keep an eye on the time.
- Try to answer every question.
- Check your answers if you have time at the end.

P40176A
©2013 Edexcel Limited.
1/1/1/1/

Turn over ▶

edexcel
advancing learning, changing lives

Using formula triangles

There will be a formula sheet in the exam, so you do not need to memorise equations, but you do need to be able to rearrange them.

If you cannot remember how to do this, you need to memorise the formula triangles given with formulae in this book. For example, $P = I \times V$ will be given in the exam paper. If you need to work out the voltage (V), cover up the V on the formula triangle. This will tell you that you need to divide P by I to get your answer.

$P = I \times V$ (given in exam)

This can be rearranged as:

$V = \dfrac{P}{I}$ or $I = \dfrac{P}{V}$

Answers

You will find some advice next to some of the answers. This is written in italics. It is not part of the mark scheme but just gives you a little more information.

Biology answers

3. Classification

1. Protoctista (1) because it is unicellular with a complex structure containing a nucleus. (1)
2. The chameleon shares more characteristics with the viper (1) because a class is a smaller group of organisms that are more similar to each other than the organisms grouped in a phylum. (1)

4. Vertebrates and invertebrates

1. (a) Any two from: have a backbone (*because they are both in the Chordata*), internal fertilisation, lay eggs, absorb oxygen from air through lungs (or any other suitable answer) (2)
   (b) Any one from: birds have feathers but reptiles do not, birds are homeotherms but reptiles are pokilotherms, birds lay hard-shelled eggs but reptiles lay leathery-shelled eggs (or any other suitable answer) (1)
2. They look similar because they have a similar lifestyle/both fast-swimming and predators. (1) Sharks are fish because they have gills and are poikilotherms (or any other suitable characteristic of fish). (1) Dolphins are mammals because they have lungs, give birth to live young, produce milk to feed the young (or any other suitable characteristic of mammals). (1)

5. Species

1. (a) able to produce young
   (b) has parents of two different species
   (c) organisms that have very similar characteristics, can interbreed and produce fertile young
2. Simple key picking out a distinguishable feature of each organism (max. 2) correctly leading to each organism. (max. 2)

6. Binomial classification

1. (a) It has several visible characteristics that are not the same as either parent species. (1) Classifying on visible characteristics only would suggest it was a different species. (1)
   (b) It was produced by interbreeding of two species. (1) It is able to breed with individuals of either parent species or other hybrids to produce fertile offspring. (1)
   (c) Because of interbreeding (1) there are individuals with a range of characteristics between the two species. (1)

7. Reasons for variety

1. Thick fur is insulation that reduces the rate of heat loss to the environment (1), white colour is camouflage against the snow and so reduces the risk of prey seeing the fox and increases the chance of the fox catching food. (1)
2. (a) variation that occurs in discrete groups, where it is not possible to have a variation that falls between the groups (1)
   (b) a characteristic that varies as a result of effects of the environment, such as weight (1)
   (c) a graph of continuous variation that shows a typical bell shape with measurements within the range being more common than measurements nearer either extreme of the range (1)

8. Evolution

1. Natural selection is the effect of the environment (e.g. climate and other organisms) on individuals so that those which are better adapted are more likely to survive than others of the same species (1) and so more likely to pass on the genes for their adaptations to

their offspring. (1) Evolution is the gradual change in characteristics of a species over time (1) as a result of natural selection. (1)
2. Species that have separated more recently through evolution will share more similar characteristics and DNA (1) than species that separated a longer time ago. (1)

9. Genes

1. (a) A chromosome is a long strand of DNA. (1)
   A gene is a small section of a chromosome/DNA that gives the instructions for producing a particular characteristic. (1)
   An allele is an alternative form of a gene. (1)
2. (a) It is heterozygous (1) because it has one dominant and one recessive allele. (1)
   (b) Purple flowers (1), because purple is dominant over white. (1) *Remember that phenotype is what it looks like, so writing down the alleles would give you no marks for this question.*

10. Explaining inheritance

1. (a) genetic diagram or Punnett square showing this information (any appropriate letter used):

		parent genotype Bb	
	parent gametes	B	b
parent genotype bb	b	Bb brown	bb black
	b	Bb brown	bb black

(**1 mark** for setting out the parent genotypes and gametes correctly, **1 mark** for completing the offspring genotypes and phenotypes correctly)
The predicted outcome of phenotypes is 50% chance of brown and 50% chance of black (could also be presented as a ratio 1 : 1, or probability 1 in 2 for both colours). (1)
   (b) The actual outcome is different to the predicted outcome because we would expect 2 black and 2 brown baby rabbits but we ended up with all black. (1)
   This is because at fertilisation it is chance which alleles are inherited. (1)

11. Genetic disorders

1. People with cystic fibrosis have thick mucus in their lungs. (1)
   This mucus is very hard for the cilia to move out of the lungs. (1)
   The mucus traps bacteria, so this means that the bacteria are not cleared out of the lung and can cause infections. (1)
   The antibiotics are used to kill the pathogenic bacteria. (1)
2. A pedigree analysis only shows phenotypes (1) so it cannot identify whether someone is heterozygous or homozygous for a dominant allele. (1)
   A blood test can be used to prove the genotype of an individual by looking at their DNA. (1)

12 and 13. Biology extended writing 1 and 2

Answers can be found on page 110.

14. Homeostasis

1. Enzymes work fastest/are most active at a particular temperature. (1)
   At lower or higher temperatures than this, they are not as active, so chemical reactions are not carried out as quickly as normal and this could cause harm to the body. (1)
2. (a) They look pink because more blood is flowing near the surface of the skin as a result of vasodilation. (1) *Either more blood flowing near skin surface or vasodilation would get this mark, but it's useful to remember the link.*

(b) This increases the rate of transfer of heat energy from the body to the environment (1) and so reduces the body temperature. (1)

3. If there is too much water in the body this means that more water is excreted in urine. (1) If there is not enough water then less water is excreted in the urine, helping to increase water levels again. (1) This is an example of negative feedback because the body works to restore the balance. (1)

15. Sensitivity

1. Sensory neurones carry electrical impulses from receptor cells to the central nervous system. (1)
Motor neurones carry nerve impulses from the central nervous system to effectors. (1)
Relay neurones link other neurones together and make up the nervous tissue of the central nervous system. (1)

2. The long axon and dendron of the sensory neurone means the cell can collect impulses from receptor cells and carry them through the body to the central nervous system/spinal cord/brain. (1)
The myelin sheath insulates the neurone from surrounding neurones and helps the electrical impulse to travel faster. (1)

16. Responding to stimuli

1. Electrical impulses in the neurones cannot cross the gap in the synapse between two neurones. (1) Neurotransmitter chemicals released from the end of a neurone when an electrical impulse arrives can cross the gap between neurones and start a new electrical impulse in the next neurone. (1)

2. Receptor cells respond to a stimulus by setting up an electrical impulse in a sensory neurone. (1) When the impulse reaches the central nervous system, another impulse is set up in a relay neurone. (1) This causes an impulse to be set up in a motor neurone which passes to an effector organ where it causes a response in the cells of the effector organ. (1)

17. Hormones

1. gland is pancreas, (1) target organ is liver (1)

2. (a) The person has just eaten a meal containing carbohydrates that have been digested in the gut. (1) Glucagon in the blood has caused a release of glucose from liver cells. (1)

   (b) The person is exercising. (1) Insulin has been released into the blood and is causing the muscle and liver cells to take in glucose from the blood. (1)

3. A change in blood glucose causes mechanisms to act that bring about the opposite change, (1) so that blood glucose concentration is maintained within a small range. (1)

18. Diabetes

1. BMI $= 100/(1.8)^2 = 100/3.24 = 30.9$ (**1 mark** for setting up equation correctly, **1 mark** for correct answer)
This is greater than 30 so he is obese. (1)

2. Exercise increases the need for glucose to provide energy, (1) so exercise can help to bring the blood glucose concentration down. (1) This means that less insulin is needed. (1)

3. There is evidence to suggest that obesity/being overweight is related to the risk of developing Type 2 diabetes. (1) As the number of people who are obese increases this means that the number of people with Type 2 diabetes is also likely to increase. (1) Controlling weight may help to prevent this. (1) However, it won't affect the number of people who develop Type 1 diabetes. (1)
Although there is no evidence that the proportion of people with Type 1 diabetes is changing much.

19. Plant hormones

1. When a shoot gets light from one side auxin from the tip moves to the shaded side of the shoot. (1) This causes cells on the shaded side to grow longer than those on the light side (1), so the shoot will curve as it grows so the tip is pointing to the light. (1)

2. Plant roots are positively gravitropic/grow downwards towards gravity. (1) Growing downwards increases the chance that the root will reach soil, which provides support, as well as water and nutrients in the soil. (1)

20. Uses of plant hormones

1. Selective weedkillers kill weed plants but not the crop plants. (1) This means that the crop plants can get more water and minerals from the soil and so grow better/bigger/produce more food for us. (1)
It's not enough here to give just the effect of the weedkillers on the different plants. You also need to explain how this benefits the crop.

2. Ripe bananas release a gas which acts as a ripening hormone on other fruit. (1)
Unripe bananas do not release this gas. (1)

21. Biology extended writing 3

Answers can be found on page 110.

22. Effects of drugs

1. Caffeine reduces reaction time. (1)
It is a stimulant (1) so you respond more quickly. (1)

2. Nicotine in tobacco smoke (1) is addictive and so difficult to do without. (1)

3. Tobacco smoke contains carbon monoxide (1), which diffuses into the blood and reduces the amount of oxygen that can be carried by red blood cells. (1) This reduces the amount of oxygen that gets to the developing foetus in the womb (1), which means that growth rate is reduced. (1)

23. The effects of alcohol

1. Alcohol is a depressant. (1) Depressants slow down your reactions. (1) This is dangerous when driving as it increases the chance of an accident. (1)

2. The studies are studying different groups of people and different aspects of the problem. (1) Since they give the same conclusion, this much more likely to be reliable than drawing a conclusion from just one of the studies. (1)

24. Ethics and transplants

1. (a) For: more organs are needed for transplants and this should help (or: so fewer people who need organs will die) (1)
Against: people have the right to state what is done with their body organs and their wishes should be respected (1)

   (b) An ethical decision is one that takes into account what is right and wrong. One person's opinion may differ from another's. (1)

2. It is difficult to judge that one argument is more right than another (1) because different people have different ideas of what is right or wrong depending on their point of view. (1)

25. Pathogens and infection

1. in water e.g. cholera (1), by air e.g. flu (1), by direct contact e.g. athlete's foot (1), in body fluids e.g. HIV (1), by vector e.g. malaria or dysentery (1)

2. Stream water may contain pathogens, such as cholera. (1) Boiling the water will kill the pathogens so you are unlikely to be infected when you drink it. (1)

3. Pathogens can get through an open cut into the blood and cause infection. (1)
A bruise does not break the skin surface, so the skin remains a barrier to pathogens. (1)

26. Antiseptics and antibiotics

1. B (1) because the area clear of bacteria is much bigger than around A (1)

2. People who are being treated with different kinds of antibiotics are likely to be closer together in a hospital than in the community. (1) So there is greater chance that bacteria already resistant to one antibiotic will come into contact with other antibiotics, which will result in the development of bacteria resistant to all those antibiotics. (1) *Other relevant answers would be acceptable.*

27. Interdependence and food webs

1. **(a)** as chemical energy in faeces and urine **(1)**
   as heat energy from respiration **(1)**
   (b) as light energy from sunlight **(1)**
2. If more energy enters the system as light energy converted to chemical energy by plants **(1)** then there will be more energy in every equivalent trophic level of a tropical food chain than in a temperate food chain. **(1)** So there may be enough energy in the tropical trophic level equivalent to the top temperate trophic level to support a higher trophic level, producing a longer food chain. **(1)**

28. Parasites and mutualists

1. A parasite is an organism that feeds on and causes harm to a host organism while living on or in the host. **(1)**
2. when two organisms both benefit from a close relationship **(1)**
3. The bacteria benefit from getting food from the plant and protection from the environment. **(1)** The plant benefits from getting nitrogen compounds/nitrates from the bacteria, which it needs to make proteins for healthy growth. **(1)**

29. Pollution

1. Eutrophication increases the rate of growth of plants and algae. **(1)** This can cause plants deeper in the water to die because they cannot get enough light for photosynthesis. **(1)** Bacteria feeding on the dead plants take oxygen from the water. **(1)** There is not enough oxygen for fish, so they die. **(1)** *Watch out: eutrophication does not kill fish directly. They die because of lack of oxygen caused by the bacteria taking more oxygen from the water.*
2. Any three from: human population is likely to increase at least for the next 50 years; **(1)** many human activities produce pollution; **(1)** so, as population increases, the amount of pollution may also increase; **(1)** we could develop ways of reducing pollution, and there may be other factors in the future that we don't know about now that might affect pollution; **(1)** so this statement is not a certainty. **(1)**

30. Pollution indicators

1. **(a)** Stonefly larvae only grow well in water with lots of oxygen **(1)** so they indicate that there is no pollution in the water. **(1)**
   (b) Blackspot fungus is killed by air pollution/sulfur dioxide in the air **(1)** so the presence of the fungus on roses shows there is no air pollution. **(1)**
2. Recycling means we don't need to use new resources. **(1)** Recycling also reduces the amount of waste we dump in landfill tips or incinerate/burn, which can also damage the environment. **(1)** *For this answer 'using fewer new resources' is not enough. You need to link this to damage to the environment to get the marks.*
3. Samples from above and below a source of pollution will show different indicator organisms present. **(1)** Organisms such as stonefly and freshwater shrimps will be replaced by organisms such as bloodworms and sludgeworms where the water is polluted. **(1)** The nearest source above stream to where the organisms have changed will be the most likely source of the pollution. **(1)**

31. The carbon cycle

1. Decomposers release carbon back into the air as carbon dioxide from respiration. **(1)**
2. Respiration releases carbon from compounds in living organisms as carbon dioxide gas into the air. **(1)** Photosynthesis takes carbon dioxide gas from the air and converts it into carbon compounds in plants. **(1)** Combustion releases carbon from compounds in fossil fuels as carbon dioxide gas into the air. **(1)**

32. The nitrogen cycle

1. by soil bacteria that convert nitrogen to ammonia to and then to nitrates **(1)**
   by lightning that converts nitrogen in the air to nitrates **(1)**

It is not enough only to mention bacteria and lightning. To get the marks, you need to explain how they increase the amount of nitrate in the soil.

2. The stubble contains proteins. **(1)** Decomposers break down these proteins and release the nitrogen compounds as ammonia into the soil. **(1)** This makes the ammonia available to nitrifying bacteria, which convert it to nitrates. **(1)** Plants that grow in this soil later will be able to take in the nitrates and use them to make proteins, which they need for healthy growth. **(1)**

33 and 34. Biology extended writing 4 and 5

Answers can be found on page 110.

Chemistry answers

35. The early atmosphere

1. The early atmosphere is thought to have been produced by volcanoes, **(1)** and volcanoes emit a lot of carbon dioxide. **(1)** There are iron compounds in very old rocks that can only form if there is no oxygen. **(1)**
2. The amount of water vapour in the atmosphere decreased when the Earth had cooled enough for liquid water to exist. **(1)** The changes in carbon dioxide and oxygen both depend on life. **(1)** so these changes could not start to happen until life had evolved. **(1)**

36. A changing atmosphere

1. The amount of water vapour decreased when the oceans condensed. **(1)** The amount of carbon dioxide decreased when life evolved and marine organisms used carbon to make shells. **(1)** The amount of oxygen increased when photosynthesising organisms released oxygen as a waste product. **(1)**
2. **(a)** Burning the trees adds carbon dioxide to the atmosphere **(1)** and farm animals such as cows release methane. **(1)**
   (b) Burning fossil fuels releases carbon dioxide into the atmosphere. **(1)**

37. Rocks and their formation

1. Sedimentary rocks are eroded more easily than igneous rocks. **(1)** This is probably because igneous rocks are made from interlocking crystals, **(1)** but sedimentary rocks are made from grains stuck together. **(1)** *Just saying that sedimentary rocks are softer is not a very good explanation.*
2. Fossils are formed when organisms die and become buried in sediment before they can rot away **(1)** and their shape is preserved when the sediments become rock. **(1)** Organisms falling into the molten rock that cools to form igneous rocks would be destroyed. **(1)**

38. Limestone and its uses

1. **(a)** Any three from: roads, buildings, glass, cement, concrete **(1 mark** for each one)
   (b) roads or building
2. Economic factors (advantages): providing more jobs, **(1)** workers at the quarry spending more money in local shops **(1)** *Providing important raw materials and helping the UK economy are economic advantages, but they are not really* local *advantages.* Environmental factors (disadvantages): Any two from: producing dust, producing noise, destroying habitats, spoiling the scenery **(1 mark** each) *Producing dust and noise could also be considered to be social disadvantages, as they affect the people living nearby.*
3. $CaCO_3 \rightarrow CaO + CO_2$ (correct reactants **(1)**, correct products **(1)**, balancing **(1)**)

39. Formulae and equations

1. **(a)** Mixture of two compounds, **(1)** because the salt does not react with the water (or because dissolving is a physical change, not a chemical change) **(1)**

(b) NaCl(aq) **(1)** *The (aq) tells us that the substance is dissolved in water, so we do not need to include H_2O.*

2. Molecules, element **(1)**
   The chlorine atoms are joined together in molecules **(1)** and both atoms in each molecule are the same, so the substance is an element. **(1)**

40. Chemical reactions

1. zinc carbonate → zinc oxide + carbon dioxide **(1)**
2. $Mg(s) + 2HCl(aq) → MgCl_2(aq) + H_2(g)$ **(1 mark** for reactants, **1 mark** for products, **1 mark** for the correct state symbols, **1 mark** for balancing the equation correctly). *All acids are solutions, so should always have (aq) as the state symbol.*
3. $Pb(NO_3)_2(aq) + 2KI (aq) → 2KNO_3(aq) + PbI_2(s)$ **(2 marks** for putting the substances on the correct sides of the equation, **1 mark** for the state symbols, **1 mark** for correct balancing)

41. Reactions of calcium compounds

1. (a) it turns cloudy (or milky) **(1)**
   (b) $Ca(OH)_2(aq) + CO_2(g) → CaCO_3(s) + H_2O(l)$ **(1 mark** for correct reactants, **1 mark** for correct products, **1 mark** for correct state symbols)
2. (a) There is fizzing (or bubbles), **(1)** steam is seen **(1)** and the calcium oxide crumbles to a white powder. **(1)**
   (b) $CaO(s) + H_2O(l) → Ca(OH)_2(s)$ **(1 mark** for correct reactants, **1 mark** for correct products, **1 mark** for correct state symbols). *Your equation would also be marked correct if you put (aq) with calcium hydroxide, as this depends on how much water is added.*

42. Chemistry extended writing 1

Answers can be found on page 110.

43. Indigestion

1. The remedy neutralises some of the acid **(1)** so the pH rises. **(1)**
2. $2HCl(aq) + Mg(OH)_2(aq) → MgCl_2(aq) + 2H_2O(l)$ **(1 mark** for substances, **1** for state symbols, **1** for correct balancing)

44. Neutralisation

1. (a) copper oxide + sulfuric acid **(1)** → copper sulfate + water **(1)**
   (b) copper carbonate + hydrochloric acid **(1)** → copper chloride + carbon dioxide + water **(1)**
2. (a) $CuO(s) + H_2SO_4(aq) → CuSO_4(aq) + H_2O(l)$ **(1 mark** for reactants, **1 mark** for products, **1** for state symbols)
   (b) $CuCO_3(s) + 2HCl(aq) → CuCl_2(aq) + H_2O(l) + CO_2(g)$ **(1 mark** for reactants, **1 mark** for products, **1 mark** for state symbols, **1 mark** for correct balancing)

45. The importance of chlorine

1. The electrolysis of sea water produces chlorine, **(1)** which is a toxic gas. **(1)** Ventilating the laboratory makes sure the gas does not build up to dangerous levels. **(1)**
2. hydrochloric acid → hydrogen + chlorine **(1)**
3. $2HCl(aq) → H_2(g) + Cl_2(g)$ **(1 mark** for H_2 and Cl_2, **1 mark** for state symbols, **1 mark** for balancing)

46. The electrolysis of water

1. water → hydrogen + oxygen **(1)**
2. hydrogen + oxygen → water **(1)**
3. $2H_2O(l) → 2H_2(g) + O_2(g)$ **(1 mark** for substances, **1 mark** for state symbols, **1 mark** for balancing)

47. Ores

1. Platinum is very unreactive **(1)** so it does not combine with other elements. **(1)**
2. (a) by heating with carbon **(1)**
   (b) tin oxide + carbon → tin + carbon dioxide **(1)**

3. $2Al_2O_3(l) → 4Al(l) + 3O_2(g)$ **(1 mark** for substances, **1 mark** for state symbols, **1 mark** for balancing) *Remember that electrolysis only works on liquids. The electrolysis is carried out at high temperatures so the aluminium formed is also liquid.*

48. Oxidation and reduction

1. Lead is more resistant to corrosion **(1)** because it is less reactive than zinc. **(1)**
2. copper + oxygen → copper oxide **(1)**
3. (a) $SnO_2 + C → Sn + CO_2$ (1 for substances, 1 for balancing) *Carbon monoxide is sometimes formed, so you would also get marks for $SnO_2 + 2C → Sn + 2CO$*
   (b) $Sn + O_2 → SnO_2$ (1 for substances, 1 for balancing) *The tin oxide formed can also be SnO, so you would also get marks for $2Sn + O_2 → 2SnO$*

49. Recycling metals

1. For some metals it costs more to collect, sort and transport them **(1)** than is saved by recycling them. **(1)**
2. Lead ore releases sulfur dioxide into the atmosphere, **(1)** but this does not happen when lead is recycled. **(1)**
   Less energy is needed to recycle lead than to extract it, **(1)** so less carbon dioxide is released from burning fossil fuels. **(1)**

50. Alloys

1. In iron, layers of atoms can slide over each other **(1)** when there is a force on the metal. **(1)**
   In steel there are atoms of different sizes, **(1)** which stop the layers sliding when they are pushed. **(1)**
2. 625 **(1)** any value between 13 and 17 **(1)** *It is actually 15 carat gold.*
3. A smart material is a material whose properties change if the conditions change. **(1)** If their shape is changed, **(1)** they resume their original shape when they are heated. **(1)**

51 and 52. Chemistry extended writing 2 and 3

Answers can be found on page 111.

53. Crude oil

1. (a) fractional distillation **(1)**
   (b) The fractions are more useful **(1)** than the crude oil mixture. **(1)**
2. C_8H_{18} **(1)**
3. Each fraction still contains hydrocarbon molecules with different numbers of carbon atoms **(1)** so the fractions are not pure substances. **(1)**

54. Crude oil fractions

1. They are similar because they are all hydrocarbons, **(1)** which release energy when they burn. **(1)**
   Any two differences from: petrol and diesel have molecules with shorter carbon chains, lower boiling points, easier to ignite, are less viscous **(2)**
2. Any two from: it is difficult to light **(1)** and it is very viscous **(1)** so it would be hard to store and send through pipes to the engine **(1)**

55. Combustion

1. Bubble the gas through limewater **(1)**; if the gas is carbon dioxide the limewater will turn milky. **(1)** *You can say milky or cloudy, or 'forms a white precipitate'.*
2. (a) propane + oxygen → carbon dioxide + water **(1)**
   (b) propane **(1)**
   (c) $C_3H_8(g) + 5O_2(g) → 3CO_2(g) + 4H_2O(l)$ **(1 mark** for substances, **1 mark** for state symbols, **1 mark** for balancing) *Water is produced as a gas in the reaction because of the high temperature, but the state symbol is given as (l) because water is a liquid at room temperature. However you would still be marked correct if you put (g) as the state symbol for water.*

56. Incomplete combustion

1. Carbon monoxide is toxic **(1)** because it reduces the amount of oxygen that the blood can carry. **(1)**

2. Three from the following: both reactions produce carbon dioxide and water; **(1)** incomplete combustion also produces carbon/soot **(1)** and carbon monoxide; **(1)** complete combustion always produces the same number of carbon dioxide and water molecules for each molecule of fuel, **(1)** but in incomplete combustion the amounts of the different products can vary. **(1)**

3. The soot shows that incomplete combustion is taking place **(1)** Carbon monoxide *may* also be forming, but the soot *does not directly* show that this is happening. **(1)**

57. Acid rain

1. Limestone is made of calcium carbonate. **(1)**
   The acid in the rain reacts with the calcium carbonate/limestone. **(1)**
   The products are washed away by the rain. **(1)**
   calcium carbonate + sulfuric acid **(1)** → calcium sulfate + carbon dioxide + water **(1)**

2. Acid rain does have a pH less than 7 **(1)** but so does normal rain **(1)** because carbon dioxide in the air dissolved in it and makes it slightly acidic. **(1)**
   Acid rain is rain that is more acidic (or has a lower pH) than normal rain. **(1)**

58. Climate change

1. **(a)** methane, **(1)** carbon dioxide, **(1)** water vapour **(1)**
   (b) carbon dioxide is increasing mainly because humans are burning more fossil fuels, **(1)** methane is increasing because of farming activities **(1)**

2. Seeding the oceans with iron **(1)** encourages microscopic life to grow, which use carbon dioxide from the atmosphere. **(1)**
   Converting carbon dioxide from power stations **(1)** into hydrocarbons to use as fuels. **(1)**

59. Biofuels

1. They are similar because they are both made from living organisms. **(1)**
   They are different because wood can be burnt directly, **(1)** but ethanol has to be manufactured **(1)** from sugar cane (or sugar beet). **(1)**

2. Biofuels are renewable **(1)** as more plants can be grown to replace the ones used as fuels. **(1)**
   They add less carbon dioxide to the atmosphere than other fuels **(1)** because the plants they were made from took carbon dioxide out of the atmosphere when they grew. **(1)**

3. Although the plants they are made from absorb carbon dioxide from the atmosphere and release the same amount when the fuel is burnt, **(1)** energy is also needed for making the fuel/growing or harvesting the crops/transporting the crops or the fuel, **(1)** so that overall some carbon dioxide is added to the atmosphere by the whole process. **(1)**

60. Choosing fuels

1. Similarities: both hydrocarbons; **(1)** both produce carbon dioxide and water when they burn **(1)**
   Differences: petrol is a liquid at normal temperatures, methane is a gas; **(1)** petrol is obtained from crude oil, methane is obtained from natural gas **(1)**

2. Petrol is a liquid, so it is easier to transport/use in an engine; **(1)** coal is a solid so it is difficult to use **(1)**

61. Alkanes and alkenes

1. **(a)**

   H—C—C—C—H with H atoms
   propane
   (1 mark for the correct number of carbons correctly joined, **1 mark** for the rest of the molecule correct)

(b) ethene

H₂C=CH₂ structure

(1 mark for correct number of carbon atoms, **1** for correct number of hydrogen atoms, **1** for joining them up correctly)

2. **(a)** They both contain two carbon atoms, **(1)** they both have hydrogen atoms joined to the carbon atoms. **(1)** *You could also say that they are both hydrocarbons for the 2nd mark.*
   (b) Ethene has a double bond between the two carbon atoms, but ethane only has a single bond. **(1)** *If you are asked to describe differences, you need to say something about both of the things you are comparing.*

3. **(a)** There would be no colour change **(1)** because ethane does not contain a double bond. **(1)** *You would also get the 2nd mark if you said that ethane is saturated, or that it does not react with bromine.*
   (b) The colour of the bromine water would change from orange to colourless, **(1)** because it reacts with unsaturated compounds (or compounds containing a double bond). **(1)**

62. Cracking

1. There is a greater demand for small molecules such as petrol than for larger molecules such as fuel oil, **(1)** but there is more fuel oil in crude oil. **(1)** Cracking allows more useful, smaller molecules **(1)** to be made from the longer, less useful ones. **(1)**

2. **(a)** The gas produced by cracking collects above the water at the top of the tube **(1)** so the gas in the tube is all ethene **(1)** (or, to collect the gas produced **(1)** because the gas is insoluble. **(1)**
   (b) The first few bubbles are mainly air **(1)** that expanded when the tube was heated. **(1)**

63. Polymerisation

1. Slippery, so food does not stick to it **(1)** and tough, so it is not damaged/scratched by cooking spoons etc. **(1)**

2. Any four from: a monomer is a small molecule **(1)** which is unsaturated/has a double bond **(1)** and monomers are either gases or liquids; **(1)** a polymer is a long molecule **(1)** made up of many monomers joined together **(1)** and polymers are solids **(1)**

3. **(a)** C_2H_3Cl **(1)** *It doesn't matter if you have the elements in a different order, as long as you have the correct number with each element.*
   (b)

   n CH₂=CHCl → —[CH₂—CHCl]—ₙ structure

   (1 mark for correct monomer, **1 mark** for correct polymer, **1 mark** for the *n*)

64. Problems with polymers

1. Burning advantages (any two advantages and any two disadvantages from): does not use up landfill sites; useful energy can be obtained from the combustion; **(1)** burning disadvantage: releases toxic gases **(1)** *Many students think the polymers give off toxic gases in landfill, but the toxic gases are only produced when the polymers are burnt.* Landfill advantage: does not release toxic gases; **(1)** landfill disadvantage: takes up space and we are running out of landfill sites **(1)**
   A disadvantage of both methods: loss of a valuable resource *There may be other advantages and disadvantages – correct answers will gain marks even if you write things that are not in the specification.*

2. Any four points from the following: Biodegradable polymers will rot. **(1)** This is an advantage because it means they will not fill up landfill sites for as long as non-biodegradable polymers. **(1)** But it may also mean that they will rot while they are in use, **(1)** so products made from them may not be as long-lasting. **(1)** People are more likely to throw them away than recycle them, **(1)** so overall more raw materials will be used in making polymer products. **(1)**

65 and 66. Chemistry extended writing 4 and 5

Answers can be found on page 111.

Physics answers

67. The Solar System

1. (a) The geocentric model has the Earth in the centre with the Sun and planets orbiting around it, **(1)** and the heliocentric model has the Sun in the centre with the Earth and planets orbiting around it. **(1)**
    (b) He discovered that Jupiter has moons orbiting around it. **(1)**
    (c) The heliocentric model said that everything in the sky moved around (orbited) the Earth **(1)**, but moons orbiting Jupiter was evidence that some things moved around (orbited) a different body. **(1)**

2. Any four from: The six planets closest to the Sun are large enough/bright enough to be seen with the naked eye, **(1)** but Uranus and Neptune are further away **(1)** and so appear dimmer/smaller. **(1)** Telescopes have improved/got more powerful over time. **(1)** Uranus and Neptune were discovered when telescopes had improved enough to make them visible. **(1)**

68. Reflection and refraction

1. (a) The distance between the centre of the lens **(1)** and the point to which parallel **(1)** rays of light are brought together. **(1)**
    (b) Any four from: Use the converging lens to focus parallel rays of light **(1)** such as from a distant scene **(1)** onto a screen. **(1)** Measure the distance between the centre of the lens and the screen. **(1)** This is the focal length of the lens. **(1)**

2. Any four points from: If the waves hit the boundary at an angle **(1)** the first part of the wave to meet the boundary will speed up **(1)** so its wavelength will increase. **(1)** The rest of the wavefront will speed up as it reaches the boundary. **(1)** This has the effect of making the direction of travel **(1)** bend away from the normal. **(1)**

69. Telescopes

1. (a) eyepiece lens **(1)**   (b) eyepiece lens **(1)**

2. Any four from: Reflecting telescopes use a lens **(1)** to magnify the light collected by the curved mirror, **(1)** so the amount of light will be reduced as it passes through the lens. **(1)**
    Refracting telescopes use two lenses, **(1)** one to collect light and one to bring it to a focus, **(1)** so more light is absorbed. **(1)**

70. Waves

1. amplitude **(1)**   wavelength **(1)**   for correct shape **(1)**

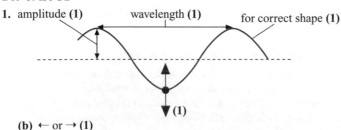

(1)

    (b) ← or → **(1)**

2. 2.2 mm **(1)** *Any answer between 2.1 and 2.4 is acceptable.*

71. Wave equations

1. $\lambda = 330$ m/s $\div 100$ Hz **(1)**
    $= 3.3$ m **(1 mark** for the number, **1 mark** for the correct unit)

2. 1 minute $= 60$ seconds **(1)**
    $x = 25$ m/s $\times 60$ s **(1)**
    $= 1500$ m
    (1 mark for the number, **1 mark** for the correct unit)

72. Beyond the visible

1. The temperature rise would be smaller **(1)** than that for violet light **(1)** because the temperature rise decreases from the red to the violet end of the visible spectrum. **(1)**

2. (a) 750 nm, **(1)** 400 THz **(1)** *Red has the longest wavelength and lowest frequency. Red light is actually a range of wavelengths and frequencies, but the ones given in the question are the only numbers you know are relevant to red light.*
    (b) about 570 nm **(1)** and 590 THz **(1)** *These values are about half way between those given in the question.*

73. The electromagnetic spectrum

1. microwave **(1)** *(1 mm = 10^{-3} m)*

2. speed $= 6 \times 10^7$ Hz $\times 5$ m **(1)**
    $= 3 \times 10^8$ m/s **(1 mark** for the number, **1 mark** for the unit)

3. (a) 3×10^8 m/s **(1)** as this is the same speed as the radio wave in question 2 **(1)** *If you got the answer to question 2 wrong, but put the same number here, give yourself the mark. In an exam you only lose a mark once for the same mistake.*
    (b) 2.25×10^8 m/s **(1)** as light slows down when it goes into glass **(1)** *See page 68.*

74. Dangers and uses

1. Any three from: X-rays can pass through some body tissues/can be used to image inside the body, **(1)** so they can be used to help to diagnose problems. **(1)** X-rays rays can cause mutations in cells, **(1)** which may lead to cancer. **(1)**

2. Any three from: Microwave ovens work by heating water inside the food **(1)** so food cooks from the inside out. **(1)** Grills and toasters emit infrared radiation, which is absorbed by the surface of food **(1)** so the food cooks from the outside. **(1)** Food cooked in a microwave does not go brown on the outside. **(1)**

75. Ionising radiation

1. Gamma rays are the higher frequency end of the electromagnetic spectrum. **(1)** The higher the frequency, the more energy is transferred. **(1)** Gamma rays have enough energy to cause ionisation but visible light with a much lower frequency does not have enough energy to cause ionisation. **(1)**

2. Ionising radiation can ionise atoms in the body. **(1)** The ions are reactive and can damage DNA. **(1)** This can lead to cancer **(1)** or mutate or kill cells. **(1)**

76. Physics extended writing 1

Answers can be found on page 111.

77. The Universe

1. Moon – 3500 km; Earth – 13 000 km; Jupiter – 143 000 km; Sun – 1.4×10^6 km; Milky Way – 1.0×10^{18} km **(2 marks** for all correct, **1 mark** for 3 correct)

2. The Moon is much closer to the Earth than the other planets are **(1)** so it is easier (or takes less time) to get there. **(1)**

78. Exploring the Universe

1. The Earth's atmosphere absorbs X-rays, **(1)** so an X-ray telescope on the ground would not detect anything (or, so the X-ray telescope must be above the atmosphere). **(1)**

2. (a) Visible light can pass through the atmosphere **(1)** so telescopes on the ground can detect it. **(1)**
    (b) So that they are above the dust and clouds that can block the view of the stars. **(1)**
    The atmosphere can also distort images and make them less clear. **(1)**

79. Alien life?

1. All life on Earth depends on liquid water. **(1)**
    Oxygen on Earth is produced by plants (or, oxygen may indicate some form of plant life). **(1)**

2. take close-up images **(1)** because an orbiter is a long way from the surface (or the lander is on the surface); **(1)** test soil samples **(1)** because it needs to be in contact with the surface to collect the soil to be tested **(1)**

3. The soil may contain microscopic organisms **(1)** or substances that are only made by living organisms. **(1)**

80. Life-cycles of stars

1. A white dwarf star has used up all its nuclear fuel **(1)** so it is no longer being heated. **(1)**

2. Gravity pulls dust and gas in a nebula together to form a star. **(1)** Gravity pulls the remains of a red giant together to form a white dwarf (or, gravity pulls the remains of a red supergiant together to form a neutron star or black hole). **(1)**

3. Supernovas occur when red supergiants reach the ends of their lives **(1)** and the Sun is not massive enough to form a supergiant. **(1)**

81. Theories about the Universe

1. The wavelength has increased. **(1)**
   Light travels at a constant speed in vacuum **(1)** so if its wavelength has increased its frequency must have decreased. **(1)** *Remember that red is the longer-wavelength end of the visible spectrum, so if light is 'red-shifted' its wavelength is longer.*

2. Any four from: The Universe is expanding, **(1)** so the galaxies are moving away from us **(1)** and the light from both of them will be red-shifted. **(1)** More distant galaxies are moving away from us faster, **(1)** so their light will have more red-shift. **(1)** The Pinwheel galaxy is further away than NGC55, so light from it will have a greater red-shift. **(1)**

82. Infrasound

1. Infrasound, ultrasound and normal sound are all longitudinal waves. **(1)** Infrasound has lower frequencies (or longer wavelengths) than normal sound. **(1)** Ultrasound has higher frequencies (or shorter wavelengths) than normal sounds. **(1)**

2. Infrasound can travel further than normal sounds. **(1)** Whales need to communicate over long distances. **(1)**

3. Many volcanoes are in remote locations, where there may be no local people to report the eruption, or where the eruption may make it impossible for the local population to report the eruption. **(1)** If there are people in the area, finding out about the eruption allows help to be sent. **(1)** Even if there are no people, learning about the eruption allows it to be studied so that scientists can learn more about how volcanoes work. **(1)**

83. Ultrasound

1. Ultrasound waves are sent into the woman's body, **(1)** they are reflected by layers of tissue, **(1)** the echoes are detected by the scanner, **(1)** a computer builds the information into a picture. **(1)**

2. $v = \frac{x}{t}$ **(1)**
   $= 280 \text{ m}/0.2 \text{ s}$ **(1)**
   $= 1400 \text{ m/s}$ **(1)**

84. Seismic waves

1. Any three from: Both types of wave are produced by earthquakes and explosions, **(1)** and travel through the Earth. **(1)** S waves are transverse and P waves are longitudinal. **(1)** P waves travel faster than S waves. **(1)**

2. Any four from: Seismic waves are detected by seismometers. **(1)** P waves travel faster than S waves, so they arrive at the seismometers first. **(1)** The greater the difference in arrival times, the further away the earthquake was, **(1)** so the distance from the earthquake to each seismometer can be worked out from the difference in the arrival times. **(1)** The distances of the earthquake from many different seismometers are used to work out the location, **(1)** at least 3 different seismometers are required.

3. Refraction happens when the speed of a wave changes. **(1)** The properties of the rocks making up the Earth change gradually with depth, **(1)** so direction of travel of the waves changes gradually as they go deeper into the Earth. **(1)**

85. Predicting earthquakes

1. **(a)** 70 km = 70 000 m **(1)**
   time = distance/speed **(1)**
   = 70 000 m/45 m/s **(1)**
   = 1555 seconds **(1)**

   (b) Scientists (or their computer systems) would take a little time to work out where the earthquake happened and so whether or not a tsunami was likely. **(1)**

2. Any two from: There are many factors that affect when an earthquake will happen **(1)** such as how strong the rocks are **(1)** or how much force has built up **(1)** and these factors are very difficult to measure accurately **(1)** or cannot be measured at all. **(1)**

86 and 87. Physics extended writing 2 and 3

Answers can be found on page 111.

88. Renewable resources

1. Any three from: Tidal power is not available all the time, but is reliable and available at predictable times, **(1)** whereas hydroelectricity is available at any time. **(1)** There are not many suitable locations for tidal barrages/tidal turbines; **(1)** there are not many suitable places for hydroelectricity either. **(1)**

2. Any four from: Wind turbines are cheaper than building geothermal plants. **(1)** There are more suitable locations for wind turbines than geothermal plants, **(1)** but each wind turbine does not produce much electricity compared to a geothermal power station **(1)** and some people think wind turbines spoil the view. **(1)** Wind power is not available all the time **(1)** but geothermal power is. **(1)** *In an exam you will lose marks if you just make lists of facts about the two things – you must compare them.*

89. Non-renewable resources

1. Nuclear-fuelled power stations are better than fossil-fuelled ones because they do not release any polluting gases. **(1)** Nuclear power stations are worse because they are more expensive to build and decommission. **(1)** *You could also have said that people worry about nuclear accidents.*

2. The advantages of fossil fuels are that the electricity they produce usually costs less than electricity produced by renewable resources **(1)** and electricity generated using fossil fuels is available all the time. **(1)** The advantages of renewable resources are that they will not run out **(1)** and they do not cause pollution by adding carbon dioxide and sulfur dioxide to the atmosphere. **(1)** *Be sure to say how fossil fuels cause pollution – just mentioning pollution will not get you the mark!*

90. Generating electricity

1. put more turns of wire on the coil, **(1)** spin the coil faster, **(1)** use stronger magnets **(1)**

2. Any two from: They need to generate large currents (or high voltages) **(1)** so they need powerful magnets **(1)** and electromagnets are stronger than permanent magnets. **(1)**

3.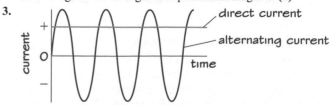

   (1 mark for each correct line) *It doesn't matter how high up your direct current is, as long as it is a straight, horizontal line.*

91. Transmitting electricity

1. Less energy is wasted than if it is sent at lower voltages **(1)** because the current is lower. **(1)**

2. Vp = (50/1000) × 400 kV **(1)**
   = 20 kV **(1)**

92. Electrical power

1. time = energy/power **(1)**
   = 1200 J/60 W **(1)**
   = 20 seconds **(1)**
2. current = power/voltage **(1)**
   = 1000 W/230 V **(1)**
   = 4.35 **(1)** A **(1)**
3. 1 minute = 60 seconds, 2 kW = 2000 W **(1)**
   energy = power × time **(1)**
   = 2000 W × 60 s **(1)**
   = 120 000 J **(1)**
   You would also get the mark if you put 120 kJ.

93. Paying for electricity

1. cost = 0.1 kW × 6 h × 14p/kW h **(1)** = 8.4p **(1)**
   When you are calculating costs, the power must always be in kW.
2. cost per kWh = cost/(power × time) **(1)**
   = 50p/(2 kW × 2 h)
   = 12.5p **(1)**

94. Reducing energy use

1. payback time = £2000/£150 per year = 13.3 years **(2)**
2. How many washes the buyer will do in a year **(1)** as the savings depend on how often the machine is used. **(1)**

95. Energy transfers

1. Wasted energy = 375 kJ − 300 kJ **(1)**
   = 75 kJ **(1)**
2. (a) gravitational potential energy **(1)** → kinetic energy **(1)**
   → sound and heat/thermal energy **(1)**
   (b) All the gravitational potential energy stored in the box when it was on the shelf **(1)** is eventually transferred to heat energy. **(1)** *Remember that sounds are movements of air particles, and when sounds dissipate (fade away) the energy is spread out into the surroundings as thermal energy.*

96. Efficiency

1. (a) 60 J − 6 J = 54 J **(1)**
   (b) efficiency = 6 J/60 J × 100% **(1)** = 10% **(1)**
2. 80% = 0.8 energy supplied = useful energy/efficiency **(1)**
   = 8 J/0.8 **(1)** = 10 J **(1)**

97. The Earth's temperature

1. The lid will reduce the power transferred to the surroundings. **(1)** There will be more power being absorbed by the water (from the heater) than being transferred from it **(1)** and the temperature of the water will rise. **(1)**
2. Any three from: Snow is white so it is a good reflector (or poor absorber) **(1)** of infrared radiation from the Sun. **(1)** Dark colours are good absorbers **(1)** and so the soot will absorb more heat from the Sun, **(1)** and the warm soot will warm the snow below it. **(1)**

98 and 99. Physics extended writing 4 and 5

Answers can be found on page 112.

Extended writing answers

Below you will find a list of points which will help you to check how well you have answered each Extended writing question. Your actual answer should be written in complete sentences, it will contain lots of detail and will link the points into a logical order. A full answer will contain most of the points listed but does not have to include all of them and may include other valid statements. You are more likely to be awarded a higher mark if you use correct scientific language and are careful with your spelling and grammar.

12. Biology extended writing 1

Binomial classification gives every organism a unique name; this name has a genus and species; an example of this would be *Homo sapiens* for humans; a correct classification allows organisms to be identified; new species can also more easily identified; similar organisms can be identified and relationships analysed; this allows areas of biodiversity to be identified; biodiversity is a measure of the number of different species in a habitat; the measure of biodiversity is more accurate if different species are easy to distinguish; conservation resources are often concentrated in these areas as they have the greatest positive effect on the largest numbers of species. *You would also get credit if you mentioned: high biodiversity is important as so many foods and medicines are derived from organisms (but you are not expected to know this).*

13. Biology extended writing 2

Data – most results in the range 160–164 cm; some results are either above or below this range; this pattern is a normal distribution; would make a bell-shaped curve.

General – range is due to continuous variation; this means both genetic and environmental factors are involved.

Genetic variation – genes carry information for how tall you can become; alleles for height are passed from parents to children; genetic factors could mean children are better/worse at making hormones or enzymes needed for growth; mutation might cause genetic disorders that inhibit growth.

Environmental variation – height depends on factors such as nutrition; other things may affect growth of foetus in pregnancy such as smoking or drinking.

21. Biology extended writing 3

Auxins are plant growth substances; they cause phototropism in shoots; auxins cause cells to elongate; this makes plant shoots grow towards the light; they also cause geotropism (positive gravitropism); this makes roots grow downwards; gibberellins cause starch to be converted into sugars; this helps seeds grow if they remain dormant for a long time before germination; gibberellins also help flowering and fruiting; artificial auxins can be used as a weedkiller; it makes some plants grow too fast; they run out of nutrients and die; plant hormones can be used to encourage root growth; this happens in rooting powder; some plant hormones encourage fruit to grow; but not seeds; this makes seedless fruit like grapes; these hormones also make the fruits larger; ethene is a hormone that helps fruit to ripen; this allows fruit to be picked and transported before it is fully ripe; it can then be ripened before being sold.

33. Biology extended writing 4

Vectors, such as flies and mosquitoes, can carry pathogens (e.g. flies carrying dysentery, mosquitoes carrying malaria parasite); contaminated food; often responsible for food poisoning through *E. coli* bacteria; direct contact; can spread the fungus that causes athlete's foot; infected blood or bodily fluids; can transmit HIV, hepatitis and other pathogens; airborne particles of virus leads to spread of influenza; water; contaminated with sewage can spread the cholera or typhoid bacteria (note that other examples not on the specification can also be counted).

34. Biology extended writing 5

The hypothesis is that the more caffeine that is consumed, the shorter the reaction time; plan should include – sensible volume of cola to be consumed; task in which reaction time can be measured; such as pressing a button in response to a flashing light; several trials to done; readings to be averaged; data from several students to be taken; experiment repeated for different amounts of cola drunk; this should include a reading with no cola drunk; data collected in a table; idea of keeping a fair test, such as same volume of cola, same stimulus; idea of measuring cola, such as a measuring cylinder; idea of some safety precautions, such as not drinking too much cola, not drinking in a science lab.

42. Chemistry extended writing 1

Sedimentary rocks can be formed when any type of rock is eroded; eroded particles are moved away by wind or water; particles are deposited; particles become compressed by more layers of sediment above them; any water in them is squashed out; the particles become sedimentary rocks; these rocks may have fossils; from the remains of plants or animals buried in the sediments; they are easier to erode than other rock types; because they are made of grains stuck together.

Igneous rocks are made from any other type of rock; this happens when rocks are heated so much that they melt; igneous rocks are formed when the molten rock cools down again; if it cools quickly it forms rocks with small crystals; if it cools slowly it forms rocks with large crystals; igneous rocks are made from interlocking crystals; so they are hard and resistant to erosion.

Metamorphic rocks are made from any other type of rock; this happens when existing rocks are heated; or subjected to great pressure; or both; new crystals form; the crystals are interlocking; so metamorphic rocks are hard and resistant to erosion.

51. Chemistry extended writing 2

Unreactive metals such as gold and silver are found as the metal itself; more reactive metals are found as ores; ores are compounds of the metal; these compounds are usually oxides; the oxide needs to be reduced to produce the metal; for some metals this can be done by heating the metal with carbon; the carbon is oxidised to carbon dioxide and the metal oxide is reduced; this works for metals less reactive than aluminium; more reactive metals must be extracted by electrolysis; the more reactive the metal, the harder it is to decompose its compounds; these metals include aluminium, magnesium, calcium, sodium and potassium *(you would only be expected to name a couple of examples)*; electrolysis is a more expensive method than heating with carbon; because of the cost of electricity.

52. Chemistry extended writing 3

The symbol means corrosive; it is one of a series of standard symbols used on containers holding hazardous substances; they warn people about the dangers of the substances; they inform people about safe ways of working with the substances; they are recognised internationally; so people who speak different languages can still understand them.

Hydrochloric acid can be neutralised by adding an alkali; acids can be neutralised by metal oxides, metal hydroxides and metal carbonates; the metal oxide and hydroxide will react to form a salt and water; the carbonate will form a salt and water and carbon dioxide.

$2HCl + CuO \rightarrow CuCl_2 + H_2O$

$HCl + NaOH \rightarrow NaCl + H_2O$

$2HCl + CaCO_3 \rightarrow CaCl_2 + H_2O + CO_2$

You could have used other examples, as long as you have balanced the equation correctly.

65. Chemistry extended writing 4

Cracking involves breaking down long hydrocarbon molecules into smaller ones; by heating; the products are shorter chain alkanes and alkenes; cracking is done because the percentages of different fractions available from crude oil are not the same as the demand for the fractions; crude oil contains more than is wanted of bitumen; fuel oil; diesel oil; and kerosene; and there is not enough petrol; and gases; to meet demand; if oil was not cracked more crude oil would have to be extracted; to meet the demand for petrol and gases; and a lot of the other fractions would be wasted; and this would cost money/reduce profits; the shorter chain alkanes produced by cracking are used as fuels; the alkenes produced can be used to make polymers; so nothing is wasted; so even though it costs money to carry out the cracking process; it is more profitable overall to do the cracking.

66. Chemistry extended writing 5

Wasting resources: plastic milk bottles are made from polymers; which are made from alkenes; alkenes are made from long chain hydrocarbons obtained from crude oil; by cracking the longer molecules; these longer molecules are less useful; and cracking also produces more useful, shorter molecules; so making plastic milk bottles is not wasting crude oil.

Advantages: plastic bottles are light; so they don't add too much to transport costs; they don't break if they are dropped.

Disadvantages: used plastic milk bottles are a problem because they fill up landfill sites; or produce toxic waste if they burn; but they can be recycled.

Evaluation: as making plastic bottles does not really waste oil; the advantages of using plastic bottles outweigh the disadvantages; as long as they are recycled where possible. *You would also get marks for saying the disadvantages outweigh the advantages, as long as you gave a sensible reason for your opinion.*

76. Physics extended writing 1

This question could be answered using a table, a set of bullet points or by writing two or three short paragraphs. The answers below talk about the similarities and then the differences. For the differences, X-rays are mentioned first and then ultraviolet, but you could also organise your answer by writing about uses first and then dangers.

They are similar because: both types travel at the same speed in a vacuum; they can both be refracted and reflected; they can both cause cancer.

They are different because they have different properties and uses.

X-rays have a higher frequency than ultraviolet, so they transfer more energy. X-rays can pass through some materials, and so they are useful for taking X-rays of people in medicine, and to 'see' inside luggage in airport security scanners. They are a form of ionising radiation, and they cause cancer by damaging DNA.

Ultraviolet is used in fluorescent lamps, to detect forged bank notes, and is also used to disinfect water. We are exposed to ultraviolet radiation any time we are in the sun, but we are only exposed to X-rays during scans in hospital. Ultraviolet radiation can also damage the eyes and cause sunburn and skin cancer.

86. Physics extended writing 2

Telescopes collect radiation coming from stars/galaxies; the atmosphere absorbs some wavelengths of electromagnetic radiation; so telescopes on the ground would not be able to detect some wavelengths; X-ray telescopes must be on satellites; ultraviolet telescopes must be on satellites; visible light can pass through the atmosphere so visible light telescopes can be on the ground; the atmosphere absorbs some of the wavelengths of infrared; warm air in the atmosphere and warm ground emits infrared; so infrared telescopes are usually on satellites to avoid interference; some microwave radiation passes through the atmosphere; so microwave telescopes can be based on the ground; from the air; most wavelengths of radio waves pass through the atmosphere; radio telescopes can be built on the ground. *You would also get credit if you mentioned that warm air in the atmosphere and warm ground emits infrared, so infrared telescopes are usually on satellites to avoid interference (but you are not expected to know this).*

Some visible light telescopes are put on satellites; because clouds can block the view of the sky; dust and air can reduce the quality of the image; so visible light telescopes in space can take sharper pictures; these telescopes also avoid light pollution by being in space.

87. Physics extended writing 3

Convection currents in the mantle cause tectonic plates to move; these move against each other at plate boundaries; they do not move smoothly, but in a series of jerks; each jerk is an earthquake; when an earthquake happens depends on many factors; such as the forces on the plates and the friction between them; these factors cannot be measured; so an earthquake cannot be predicted. Seismometers only detect the earthquake once it has happened, so they cannot be used to give a warning.

Tsunamis are caused by earthquakes on the sea floor; when an earthquake happens seismic waves spread out from it; scientists can use seismometer data to determine where an earthquake happened; if it was beneath the sea there may be a tsunami; sea-floor pressure sensors can also detect the tsunami wave; it takes time for the wave to travel from its source to the nearest land; so people on land may get several hours warning.

98. Physics extended writing 4

Both are renewable resources; neither of them produce atmospheric pollution; hydroelectricity is available at any time; if it is not used, the gravitational potential energy stored in the water will be available later; tidal power is not available all the time but is available at predictable times; tidal power involves building barrages across river estuaries, and hydroelectricity involves building dams to create reservoirs in hills; there are not many suitable sites for either in the UK (although there are more possibilities in Scotland and Wales than in England); both are large, costly projects to undertake; both can affect habitats when they are built, through noise and disturbance during construction; tidal barrages can permanently affect habitats; reservoirs for hydroelectricity can destroy habitats, although an alternative lake habitat is created.

99. Physics extended writing 5

Measure the current in the appliance and the voltage across the appliance; change the output voltage on the power supply and measure the current and voltage again; do this for at least five different output voltages; mention some safety precautions such as not touching hot components or switching off before changing the circuit; calculate the power in watts for each setting by multiplying the current in A and voltage in V; plot the results on a graph; with voltage along the bottom and power up the side; draw a line of best fit through the points; if the line goes up from left to right then the hypothesis is correct; if the points form a straight line then the power is proportional to the voltage; make it a fair test by keeping the circuit and appliance the same and only changing the setting on the power supply; repeat the whole investigation with other appliances to see if the same conclusion applies to different kinds of appliance.

Published by Pearson Education Limited, a company incorporated in England and Wales, having its registered office at Edinburgh Gate, Harlow, Essex, CM20 2JE. Registered company number: 872828

www.pearsonschoolsandfecolleges.co.uk

Text © Pearson Education Limited 2012
Edited by Judith Head and Florence Production Ltd
Typeset by Tech-Set Ltd, Gateshead
Original illustrations © Pearson Education Limited 2012

The rights of Penny Johnson, Sue Kearsey and Damian Riddle to be identified as authors of this work have been asserted by them in accordance with the Copyright, Designs and Patents Act 1988.

First published 2012

16 15 14 13
10 9 8 7 6 5 4 3 2

British Library Cataloguing in Publication Data
A catalogue record for this book is available from the British Library

ISBN 978 144 690261 5

Acknowledgements
The author and publisher would like to thank the following individuals and organisations for permission to reproduce copyright material:

Biology
Figures
Figure 2.5/2 'Relationship between BMI and Type 2 Diabetes', 2006, http://www.cutthewaist.com/impact.html. Reproduced by permission of Cut the Waist Ltd; Figure 2.7/2 "1-Methylcyclopropene delays tomato fruit ripening", *Horticultura Brasileira*, Vol 20 (4) (MorettiI, C.L., Araújo, A.L., Marouelli, W.A., and Silva, W.L.C., 2002), The Brazilian Society for Horticultural Science. Licensed under a Creative Commons Attribution License; Figure 3.2/2 'Relative risk of an accident based on blood alcohol levels' by James Heilman, MD, http://en.wikipedia.org/wiki/File:Relative_risk_of_an_accident_based_on_blood_alcohol_levels_.png. Licensed under the Creative Commons Attribution-Share Alike 3.0 Unported license; Figure 3.2/3 adapted from "Relationship between Dietary Beef, Fat, and Pork and Alcoholic Cirrhosis", *International Journal of Environmental Research and Public Health* Vol 6 (9), pp.2417-2425 (Bridges, F.S., 2009), copyright © 2009 by the authors. Licensee Molecular Diversity Preservation International, Basel, Switzerland. Licensed under the terms and conditions of the Creative Commons Attribution license.

Tables
Table 3.2 'Risk Chart for Men (current and never smokers)' in "Risk of Death by Age, Sex, and Smoking Status in the United States: Putting Health Risks in Context", *Journal of the National Cancer Institute (JNCI)*, Vol 100, Issue 12, pp.845-853 (Woloshin, S., Schwartz, L.M. and Welch, H.G. 2008), copyright © 2008, Oxford University Press.

Chemistry
Figures
Figure 5.6/1 'Global temperature increase since 1850', www.climatechoices.org.uk/pages/cchange3.htm. Reproduced by permission of Practical Action.

Every effort has been made to contact copyright holders of material reproduced in this book. Any omissions will be rectified in subsequent printings if notice is given to the publishers.

Disclaimer
This material has been published on behalf of Edexcel and offers high-quality support for the delivery of Edexcel qualifications.

This does not mean that the material is essential to achieve any Edexcel qualification, nor does it mean that it is the only suitable material available to support any Edexcel qualification. Material from this publication will not be used verbatim in any examination or assessment set by Edexcel. Any resource lists produced by Edexcel shall include this and other appropriate resources.

In the writing of this book, no Edexcel examiners authored sections relevant to examination papers for which they have responsibility.

Copies of official specifications for all Edexcel qualifications may be found on the Edexcel website: www.edexcel.com